The Art of Creative Math

Complete Solutions

Full worked solutions to every practice-ladder problem

Dr Hui Quan

ASCENT MATH PRESS

The Art of Creative Math: Complete Solutions.

Published by Ascent Math Press. This is the companion solutions volume to *The Art of Creative Math: Eighteen Ideas That Make Hard Problems Easy*. It contains complete worked solutions to every problem in the eighteen practice ladders — the Launch, Core, Stretch, The Flash, and Proof Lab tiers. (The Flash problems are not timed drills; they are short problems meant to unlock through one central idea.) The Reflect prompts are deliberately omitted, as they ask for the reader's own words rather than a fixed answer.

First edition.

How to use these solutions

A solution you read before you have struggled is a solution wasted. The single most important habit with this volume is to *earn* each page: attempt the problem first, get stuck honestly, consult the brief nudge in the main book's answer key, attempt again — and only then turn here.

Each solution is written to model not just *what* the answer is but *how the idea found it*. Watch for the recurring shape: a sentence or two that names the key move (in coral), the working that carries it out, and a boxed final answer where the problem calls for a number. When a solution says "the key idea," that phrase is doing the real teaching — it is the transferable part, the thing you want to recognize next time on a problem that looks different but rhymes.

A few solutions to proof problems are written at the level of a clear sketch rather than a fully formalized argument, especially where the complete proof would run long (the harder existence proofs, for instance). In those cases the goal is to make the central mechanism unmistakable and leave the routine verification to you — which is itself good practice. Where a problem as posed was open-ended or admitted several valid approaches, the solution presents one clean route and notes that others exist.

Finally: if your solution differs from the one here but reaches the same answer by sound reasoning, you have not made a mistake — you have found another window onto the same object, which is the entire spirit of the book. Compare the two. The comparison is often where the deepest understanding lives.

Double Counting

Full worked solutions. Each one names the two things being counted, then shows both counts and sets them equal.

Launch

Problem 1. Add up every three-digit number that uses each of $1, 2, 3$ exactly once.

SOLUTION. There are $3! = 6$ such numbers. Rather than list them, count the *total contribution of each place value*. Fix the digit 1 in the hundreds place; the other two digits fill the remaining places in $2! = 2$ ways. So each digit appears in each of the three places exactly 2 times. The sum contributed by any one place is therefore $2(1 + 2 + 3) = 12$ times that place's value, and across the hundreds, tens, and ones places:

$$12 \cdot 100 + 12 \cdot 10 + 12 \cdot 1 = 12 \cdot 111 = 1332.$$

1332

Problem 2. A graph has 7 vertices, each of degree 4. How many edges? Could every vertex instead have degree 3?

SOLUTION. The handshake lemma says $\sum_v \deg(v) = 2E$, because each edge contributes 2 to the degree total (one per endpoint). With seven vertices of degree 4,

$$2E = 7 \cdot 4 = 28 \quad \Longrightarrow \quad E = 14.$$

For all degrees equal to 3, we would need $2E = 7 \cdot 3 = 21$. But $2E$ is always even, and 21 is odd — a contradiction. **So 7 vertices of degree 3 is impossible.** (This is the same parity obstruction that makes a 7-person "everyone shakes exactly 3 hands" party impossible.)

$E = 14$; **all-degree-3 impossible**

Core

Problem 3. Prove $1 + 2 + \cdots + (n-1) = \binom{n}{2}$ by counting the edges of K_n two ways.

SOLUTION. Consider the complete graph K_n, in which all $\binom{n}{2}$ pairs of the n vertices are joined by an edge.

Count one: by definition, choosing an unordered pair of vertices gives $\binom{n}{2}$ edges.

Count two: label the vertices $1, 2, \dots, n$ and tally each edge by its *smaller* endpoint. Vertex 1 is the smaller endpoint of edges to $2, 3, \dots, n$, which is $n-1$ edges. Vertex 2 is the smaller endpoint of edges to $3, \dots, n$: that's $n-2$ edges. Continuing, vertex k contributes $n-k$ edges, and vertex n contributes 0. Summing,

$$(n-1) + (n-2) + \cdots + 1 + 0 = 1 + 2 + \cdots + (n-1).$$

Both counts tally the same edges, so they are equal: $1 + 2 + \cdots + (n-1) = \binom{n}{2}$. ■

Problem 4. Prove the hockey-stick identity $\sum_{i=r}^{n} \binom{i}{r} = \binom{n+1}{r+1}$ by classifying $(r+1)$-subsets by their largest element.

SOLUTION. The right side $\binom{n+1}{r+1}$ counts the $(r+1)$-element subsets of $\{1, 2, \dots, n+1\}$.

Classify by the largest element. Every such subset has a largest element; call it $m+1$, where m ranges over the possible values. If the largest element is $m+1$, the remaining r elements are chosen freely from $\{1, 2, \dots, m\}$, which can be done in $\binom{m}{r}$ ways. As the largest element runs from $r+1$ (its smallest possible value, since we need r elements below it) up to $n+1$, the index m runs from r to n. Summing the disjoint cases,

$$\binom{n+1}{r+1} = \sum_{m=r}^{n} \binom{m}{r} = \binom{r}{r} + \binom{r+1}{r} + \cdots + \binom{n}{r}.$$

■

Problem 5. Evaluate $\sum_{k=2}^{n} \binom{n}{k}\binom{k}{2}$ by choosing a committee, a captain, and a co-captain; then adapt the multi-role idea (captain *and* secretary) for $\sum_k k^2 \binom{n}{k}$.

SOLUTION. First sum. Count the ways to choose a committee of any size from n people and designate two of its members as *captain* and *co-captain* (an unordered pair of distinct leaders).

Counting by committee size first: a committee of size k can be chosen in $\binom{n}{k}$ ways, and its two leaders in $\binom{k}{2}$ ways, giving $\sum_k \binom{n}{k}\binom{k}{2}$. Counting leaders-first: choose the pair of leaders from all n people ($\binom{n}{2}$ ways), then let each of the remaining $n-2$ people be independently in or out of the committee (2^{n-2} ways). Equate:

$$\sum_{k=2}^{n} \binom{n}{k}\binom{k}{2} = \binom{n}{2} 2^{n-2}.$$

(Check at $n = 5$: $\binom{5}{2} \cdot 2^3 = 10 \cdot 8 = 80$.)

Second sum. Now choose a committee and designate a captain *and* a secretary, where one person *may hold both roles*. Counting by size gives $\sum_k k^2\binom{n}{k}$ (choose the committee, then the captain in k ways and the secretary in k ways).

Counting by roles, split into two cases. If captain and secretary are the *same* person: pick that person (n ways), the rest free (2^{n-1}), giving $n\,2^{n-1}$. If they are *different*: pick the ordered pair of distinct officers ($n(n-1)$ ways), the remaining $n-2$ people free (2^{n-2}), giving $n(n-1)2^{n-2}$. Adding,

$$\sum_k k^2\binom{n}{k} = n\,2^{n-1} + n(n-1)2^{n-2} = n\,2^{n-2}(2+(n-1)) = n(n+1)2^{n-2}.$$

(Check at $n=5$: $5\cdot 6\cdot 2^3 = 240$.) ■

Stretch

Problem 6. A convex polyhedron has 12 faces, each a pentagon or a hexagon, and exactly three faces meet at every vertex. Use double counting and Euler's formula to find how many faces are pentagons.

SOLUTION. Double count the edges, two different ways. Let p be the number of pentagonal faces and h the number of hexagonal faces, so the face count is

$$F = p + h = 12.$$

Edges through faces. Each pentagon borders 5 edges and each hexagon borders 6, accounting for $5p+6h$ edge-borders. But every edge lies between exactly *two* faces, so each edge is counted twice:

$$2E = 5p + 6h.$$

Edges through vertices. Exactly three faces meet at each vertex, and three faces meeting means three edges meeting, so each vertex is an endpoint of 3 edges, giving $3V$ edge-ends. Every edge has exactly *two* ends, so

$$3V = 2E \quad\Longrightarrow\quad V = \tfrac{2E}{3}.$$

Feed Euler's formula. Write $S = 5p+6h = 2E$, so $E = \frac{S}{2}$ and $V = \frac{2E}{3} = \frac{S}{3}$. Euler's $V - E + F = 2$ becomes

$$\frac{S}{3} - \frac{S}{2} + 12 = 2 \quad\Longrightarrow\quad -\frac{S}{6} = -10 \quad\Longrightarrow\quad S = 60.$$

So $5p+6h=60$. Together with $p+h=12$ (multiply by 6: $6p+6h=72$ and subtract):

$$(6p+6h) - (5p+6h) = 72 - 60 \quad\Longrightarrow\quad p = 12.$$

Hence **all** 12 **faces are pentagons** ($h=0$) — the polyhedron is forced to be a regular dodecahedron. The double count of edges pins down the pentagon count exactly. ■

12 **pentagons (so** $h=0$**: a dodecahedron)**

Problem 7. Prove Vandermonde's identity $\sum_k \binom{m}{k}\binom{n}{r-k} = \binom{m+n}{r}$ by a committee argument.

SOLUTION. Form a committee of r people drawn from a pool of m seniors and n juniors — $m+n$ people in all.

Count one: directly, $\binom{m+n}{r}$.

Count two: classify the committee by how many seniors it contains. If it has exactly k seniors, those are chosen in $\binom{m}{k}$ ways and the remaining $r-k$ members are juniors, chosen in $\binom{n}{r-k}$ ways. Summing over all valid k,

$$\binom{m+n}{r} = \sum_k \binom{m}{k}\binom{n}{r-k}.$$

■

The Flash

Problem 8. In a tournament of n players with no ties, the total number of wins equals what, and why must it equal the number of games?

SOLUTION. Every pair of players meets exactly once, so the number of games is $\binom{n}{2}$. Each game produces exactly one win (no ties), so counting (game, its winner) incidences gives the total number of wins on one side and one-per-game on the other. Hence **total wins** $= \binom{n}{2} =$ **number of games.**

$\binom{n}{2}$

Problem 9. A $3 \times 3 \times 3$ cube is built from unit cubes. How many unit faces are visible on the surface?

SOLUTION. The big cube has 6 faces, and each face shows a 3×3 grid of unit-cube faces, i.e. 9 unit faces. Counting (big face, unit face on it) incidences,

$$6 \times 9 = 54.$$

54 visible unit faces

Proof Lab

Problem 10. In any finite collection of clubs where every two clubs share exactly one member, prove $\sum_p \binom{d_p}{2} = \binom{c}{2}$ (here c is the number of clubs and d_p the number of clubs containing person p), and state a consequence.

SOLUTION. Count pairs of clubs {club A, club B} two ways.

Count one: there are $\binom{c}{2}$ pairs of clubs, and by hypothesis each pair shares exactly one member, so each pair is counted once.

Count two: tally each club-pair by the member they share. A person p who belongs to d_p clubs is the shared member of exactly $\binom{d_p}{2}$ club-pairs (any two of their clubs share *them*, and by the "exactly one" hypothesis share no one else). Summing over people,

$$\sum_p \binom{d_p}{2} = \binom{c}{2}.$$

Consequence. This identity is one ingredient in *Fisher-type* inequalities; with additional uniformity assumptions, one can deduce that the number of clubs cannot exceed the number of people. ■

Problem 11. Prove there is no regular polyhedron whose faces are hexagons.

SOLUTION. For a regular polyhedron with p-gon faces, q meeting at each vertex, the count $pF = 2E = qV$ together with Euler's $V - E + F = 2$ forces the inequality $\frac{1}{p} + \frac{1}{q} > \frac{1}{2}$. Setting $p = 6$ (hexagonal faces),

$$\frac{1}{6} + \frac{1}{q} > \frac{1}{2} \quad \Longrightarrow \quad \frac{1}{q} > \frac{1}{2} - \frac{1}{6} = \frac{1}{3} \quad \Longrightarrow \quad q < 3.$$

But at least $q = 3$ faces must meet at every vertex of a genuine solid, so no such polyhedron exists. **Hexagons can only tile the flat plane (the honeycomb), never close into a solid.** ■

Too Many Pigeons

Full worked solutions. The craft is always the same: name the pigeons, name the holes, and count.

Launch

Problem 1. A drawer holds socks of 3 colors. How many must you grab to be sure of a matching pair?

SOLUTION. The *holes* are the 3 colors; the *pigeons* are the socks you grab. With 3 socks you might draw one of each color — no pair. But a 4th sock must repeat one of the three colors, forcing a match. **Four socks suffice, and three do not.**

4 socks

Problem 2. Show that among any 13 people, two share a birth month.

SOLUTION. There are only 12 months (holes) and 13 people (pigeons). Since $13 > 12$, by pigeonhole at least two people fall in the same month. ∎

Core

Problem 3. Among any $n + 1$ numbers from $\{1, 2, \ldots, 2n\}$, prove two are consecutive (hence coprime).

SOLUTION. Choose the boxes. Partition $\{1, \ldots, 2n\}$ into the n consecutive pairs

$$\{1, 2\},\ \{3, 4\},\ \{5, 6\},\ \ldots,\ \{2n - 1, 2n\}.$$

There are exactly n such pairs. Choosing $n + 1$ numbers (pigeons) into n boxes forces two chosen numbers into the same box — and the two members of any box are consecutive integers. Consecutive integers differ by 1, so any common divisor divides 1; they are coprime. ∎

Problem 4. Among any $n+1$ numbers from $\{1,2,\dots,2n\}$, prove one divides another.

SOLUTION. The clever boxes: odd parts. Write each chosen number uniquely as $2^a \cdot m$ with m odd; call m its *odd part*. The odd numbers in $\{1,\dots,2n\}$ are $1,3,5,\dots,2n-1$ — exactly n of them, so there are only n possible odd parts (the holes). With $n+1$ numbers, two of them, say $2^a m$ and $2^b m$, share the same odd part m. If $a<b$, then $2^a m \mid 2^b m$. ■

Problem 5. Prove some multiple of 2025 has only 0s and 1s in its decimal digits.

SOLUTION. The boxes: remainders mod 2025. Consider the 2026 numbers

$$1,\ 11,\ 111,\ \dots,\ \underbrace{11\cdots1}_{2026\text{ ones}},$$

each written with only the digit 1. Their remainders upon division by 2025 lie in $\{0,1,\dots,2024\}$ — only 2025 possible values. With 2026 numbers, two of them, say the repunits with $i<j$ ones, share a remainder. Their difference is divisible by 2025, and that difference is

$$\underbrace{1\cdots1}_{j} - \underbrace{1\cdots1}_{i} = \underbrace{1\cdots1}_{j-i}\underbrace{0\cdots0}_{i},$$

a number written with only 1s and 0s. ■

Stretch

Problem 6. Six people at a party: prove three are mutual friends or three are mutual strangers.

SOLUTION. Fix one person and apply pigeonhole to their five edges. Call one person P. Each of the other five people is either a friend or a stranger of P — two categories, five people, so by pigeonhole at least three fall in one category. Say P is friends with three people A,B,C (the stranger case is symmetric).

Now look at the three pairs among A,B,C. If any pair, say AB, are friends, then P,A,B are three mutual friends — done. If *none* of the pairs are friends, then A,B,C are three mutual strangers — also done. Either way the required triangle exists. ■

Problem 7. Prove any sequence of n^2+1 distinct reals has a monotone subsequence of length $n+1$; then show sharpness for $n=3$ by exhibiting a length-9 sequence with no monotone subsequence of length 4.

SOLUTION. Label each term with a pair, then pigeonhole the pairs. For each position i, let a_i be the length of the longest *increasing* subsequence ending at i, and d_i the longest *decreasing* subsequence ending at i. Suppose, for contradiction, that no monotone subsequence has length $n+1$; then every a_i and d_i lies in $\{1,2,\dots,n\}$, so each pair (a_i,d_i) has at most n^2 possible values.

With n^2+1 terms (pigeons) and only n^2 possible pairs (holes), two positions $i<j$ share the same pair $(a_i, d_i) = (a_j, d_j)$. But the terms are distinct: if $x_i < x_j$ then $a_j \geq a_i + 1$, and if $x_i > x_j$ then $d_j \geq d_i + 1$ — either way the pairs differ, a contradiction. Hence a monotone subsequence of length $n+1$ exists.

Sharpness for $n = 3$. The blocked staircase $3, 2, 1, 6, 5, 4, 9, 8, 7$ (length $9 = 3^2$) has no monotone subsequence of length 4: any *increasing* subsequence takes at most one term from each descending block $(3,2,1), (6,5,4), (9,8,7)$, so its length is at most 3; any *decreasing* subsequence must stay inside a single block, so its length is at most 3 too. Thus length 9 avoids a monotone subsequence of length 4, proving the threshold $10 = 3^2+1$ is sharp. ■

The Flash

Problem 8. Seventeen points inside an equilateral triangle of side 1: must two be within $\frac{1}{4}$?

SOLUTION. The boxes: 16 small triangles. Partition the triangle into 16 small equilateral triangles of side $\frac{1}{4}$ (divide each side into 4 and connect). Seventeen points among 16 small triangles force two points into the same small triangle. Two points in a triangle of side $\frac{1}{4}$ are at most $\frac{1}{4}$ apart, so yes. **Two points must be within $\frac{1}{4}$.**

Yes

Problem 9. Pick 51 numbers from $\{1, \dots, 100\}$: must two sum to 101?

SOLUTION. The boxes: complementary pairs. Partition $\{1, \dots, 100\}$ into the 50 pairs summing to 101:

$$\{1, 100\}, \ \{2, 99\}, \ \dots, \ \{50, 51\}.$$

Choosing 51 numbers among 50 pairs forces two into the same pair, and that pair sums to 101. **Yes.**

Yes

Proof Lab

Problem 10. Prove any 10 distinct two-digit numbers contain two disjoint subsets with equal sum.

SOLUTION. Compare counts of subsets and possible sums. A set of 10 numbers has $2^{10} = 1024$ subsets. Each subset's sum is at most $90+91+\cdots+99 < 1000$ (the ten largest two-digit numbers), and at least 0, so there are fewer than 1000 possible sums. With 1024 subsets (pigeons) and under 1000 sums (holes), two distinct subsets A and B share a sum.

If A and B overlap, remove their common elements from both: the removed amount is equal on each side, so the resulting sets $A \setminus B$ and $B \setminus A$ still have equal sums, are disjoint, and are nonempty (since $A \neq B$). ■

What Can't Be Undone

Full worked solutions. To prove something impossible, find the quantity the moves can never change.

Launch

Problem 1. A frog jumps ± 1 each time. Can it return to start after exactly 7 jumps?

SOLUTION. The invariant: parity of position. Each jump changes the frog's coordinate by ± 1, so it flips the parity of the position. Starting at 0 (even), after an *odd* number of jumps the position is odd, hence nonzero. Since 7 is odd, the frog cannot be back at 0. **No.**

No

Problem 2. A light switch starts OFF. After 100 toggles, is it ON or OFF?

SOLUTION. Each toggle flips the state, so the state depends only on the parity of the number of toggles. Starting OFF, an even number of flips returns it to OFF. Since 100 is even, the switch is **OFF.**

OFF

Core

Problem 3. Numbers 1 through 7 are on a board; replacing a, b with $a+b-1$ leaves one number after six steps. What is it?

SOLUTION. Find the invariant. Each move removes two numbers a, b and inserts $a+b-1$, so the sum of all numbers *decreases by exactly* 1 (we lose $a+b$, gain $a+b-1$), while the *count* of numbers also decreases by 1. The quantity

$$(\text{sum of numbers}) - (\text{count of numbers})$$

is therefore unchanged by every move. Initially the sum is $1 + 2 + \cdots + 7 = 28$ and the count is 7, so this invariant equals $28 - 7 = 21$. At the end one number x remains (count 1), so $x - 1 = 21$, giving $x = 22$. **The last number is 22, regardless of the order of moves.**

22

Problem 4. On a 5×5 board, can a rook visit every square once and return to start (one square per step)?

SOLUTION. The invariant: color, via checkerboard. Color the board like a checkerboard. A rook moving one square at a time alternates colors with each step. A closed tour visiting all 25 squares makes 25 steps and returns to start, so the colors must alternate and close up — which requires an *even* number of squares (equal counts of each color in the cycle). But 25 is odd, and a 5×5 board has 13 of one color and 12 of the other. No closed tour alternating colors can cover an odd number of squares. **Impossible.**

No

Problem 5. A jar holds 7 black and 8 white stones. Remove two: same color → add a black; different colors → add a white. What color is the last stone?

SOLUTION. The invariant: parity of the white count. Let W be the number of white stones, and check how each kind of move changes it:

- *Two blacks* removed, one black added: no white touched, W unchanged.
- *Two whites* removed, one black added: W decreases by 2.
- *One of each* removed, one white added: one white leaves and one white returns, so W is unchanged.

In every case W changes by an *even* amount, so its parity never changes. Initially $W = 8$ is even, so W stays even for the entire process. The final lone stone has $W \in \{0, 1\}$; evenness forces $W = 0$, so that stone is *black*. **Black — and the 7 black stones never mattered to the argument.**

Black

Stretch

Problem 6. In the 15-puzzle, 14 and 15 are swapped. Prove this is unreachable from solved.

SOLUTION. The invariant: permutation parity plus blank position. Model a configuration by the permutation σ of the 15 tiles together with the blank's square. Each slide of a tile into the blank is a single transposition of σ (it swaps a tile with the blank), and it moves the blank by one square — changing the "taxicab parity" of the blank's position. The combined quantity

$$\operatorname{sgn}(\sigma) \times (-1)^{\text{(blank's row+column distance from home)}}$$

is invariant under every legal slide (each slide flips both factors, so their product is preserved). In the solved board this invariant is $+1$. Swapping just 14 and 15 is a single transposition (odd permutation, $\operatorname{sgn} = -1$) while the blank is in its home corner (even displacement, factor $+1$), giving invariant $-1 \neq +1$. Since the invariant cannot change, the swapped position is unreachable. ∎

Problem 7. A 3×3 grid of ± 1 starts with a single -1 at the center. A move flips a whole row or column. Prove all $+1$ is unreachable.

SOLUTION. Track how many times each cell is flipped. Suppose row i is flipped r_i times in all and column j is flipped c_j times; only the parities matter. Cell (i,j) ends with sign (initial) $\cdot (-1)^{r_i+c_j}$. To finish at all $+1$ we need, for each cell,

$$(-1)^{r_i+c_j} = (\text{initial sign of } (i,j))^{-1}.$$

The center $(2,2)$ starts at -1, so we need $r_2 + c_2$ *odd*; every other cell starts at $+1$, so we need $r_i + c_j$ *even*.

Force a contradiction (mod 2). From the cells $(2,j)$ with $j \neq 2$: $r_2 + c_j$ even, while $r_2 + c_2$ is odd, so $c_j \not\equiv c_2 \pmod 2$ for every $j \neq 2$. From the cells $(i,2)$ with $i \neq 2$: $r_i + c_2$ even, so $r_i \equiv c_2$. Now take a corner (i,j) with $i,j \neq 2$. We require $r_i + c_j$ even, but $r_i \equiv c_2$ and $c_j \equiv c_2 + 1$, so

$$r_i + c_j \equiv c_2 + (c_2 + 1) = 1 \pmod 2,$$

which is *odd* — a contradiction. No assignment of flips can satisfy all nine requirements, so all $+1$ is unreachable. ∎

Problem 8. Among n people, some pairs are enemies. Prove everyone can be split between two rooms so each person has at most as many enemies in their own room as in the other.

SOLUTION. A monovariant that counts same-room enmity. Split the people any way to start, and let D be the number of enemy pairs who share a room. Since D is a non-negative integer, it cannot decrease forever. Suppose some person P currently has *more* enemies in their own room than across the hall. Move P to the other room. The only enemy pairs whose status changes are those involving P: the ones that were same-room become split, and the ones that were split become same-room. Because P had *more* enemies in the old room than the new, this move *strictly decreases* D.

So as long as any person is "unhappy" (more enemies at home than away), a move strictly lowers D; since D can drop only finitely often, the process must halt. When it halts, *no* person has more enemies in their own room than across — which is exactly the required seating. **The minimum of a monovariant is forced to be a stable, fair seating.** ∎

Problem 9. When a 5×5 board is covered by eight straight 1×3 tiles leaving one cell empty, prove the empty cell must be the center.

SOLUTION. Index the cells (r,c) with $r,c \in \{1,\dots,5\}$.

First coloring — necessary, not sufficient. Color (r,c) by $(r+c) \bmod 3$. A straight tromino, horizontal or vertical, covers three *consecutive* values of $r+c$, hence one cell of each color $0,1,2$. Counting the board, color 0 (where $r+c \equiv 0$) occurs 9 times and colors $1,2$ occur 8 times each. Eight tiles remove eight of each color, so the hole must lie in the surplus class $r+c \equiv 0 \pmod 3$ — nine candidate cells. But this condition is only *necessary*: the coloring alone cannot say which of the nine actually occurs.

A second coloring pins it down. Now color by $(r-c) \bmod 3$. Straight trominoes again cover one of each color (three consecutive values of $r-c$), and the same count makes the surplus class $r-c \equiv 0 \pmod 3$. So the hole must satisfy *both* $r+c \equiv 0$ and $r-c \equiv 0 \pmod 3$. Adding and subtracting, $2r \equiv 0$ and $2c \equiv 0$, i.e. $r \equiv 0$ and $c \equiv 0 \pmod 3$. Within $\{1,\dots,5\}$ the only such coordinate is 3, so the hole must be the center $(3,3)$.

And the center really works. An explicit tiling leaving $(3,3)$ empty: horizontals $(1,1)(1,2)(1,3)$, $(2,1)(2,2)(2,3)$, $(4,3)(4,4)(4,5)$, $(5,3)(5,4)(5,5)$; verticals $(1,4)(2,4)(3,4)$, $(1,5)(2,5)(3,5)$, $(3,1)(4,1)(5,1)$, $(3,2)(4,2)(5,2)$. These eight straight trominoes cover all 24 cells except the center. **So the uncovered square must be the center — and can be.** ■

The center $(3,3)$

The Flash

Problem 10. A knight makes 13 moves. Can it end on a square of the same color it started on?

SOLUTION. A knight always moves to a square of the opposite color. After 13 moves (odd), it has changed color an odd number of times, so it lands on the *opposite* color. **No.**

No

Problem 11. From $(0,0)$, each move adds $(\pm 1, \pm 1)$ (signs your choice). Can you reach $(0,1)$?

SOLUTION. Parity of the coordinate sum. Each move adds ± 1 to x and ± 1 to y, so it changes $x+y$ by $+2$, 0, or -2 — always an *even* amount. Hence the parity of $x+y$ never changes. It starts at $0+0=0$ (even), so $x+y$ is even after every move. But $(0,1)$ has $x+y=1$, which is odd. **Unreachable.** ■

No

Proof Lab

Problem 12. $n \geq 3$ lamps in a circle, all off; a move toggles a lamp and its two neighbors. For which n can all be turned on?

SOLUTION. Count flips per lamp, mod 2. Toggling lamp i flips lamps $i-1, i, i+1$. Only the parity of how often each move is used matters, so a strategy amounts to choosing a subset of lamps to toggle: let $x_i \in \{0,1\}$ record whether lamp i is toggled. Then lamp j is flipped $x_{j-1} + x_j + x_{j+1}$ times, and to go from all-off to all-on we need this to be *odd* for every j.

A construction that always works. Toggle *every* lamp once, i.e. $x_i = 1$ for all i. Then each lamp j is flipped

$$x_{j-1} + x_j + x_{j+1} = 1 + 1 + 1 = 3$$

times — an odd number — so every lamp turns on. This succeeds for *every* $n \geq 3$ (where each lamp has two distinct neighbors). **All lamps can be turned on for every $n \geq 3$** (toggle them all). ■

Problem 13. Integers on a board; a move replaces a, b (with $a \geq b+2$) by $a-1$ and $b+1$. Prove the process must terminate, via a monovariant.

SOLUTION. The monovariant: sum of squares — an *integer*. Every value stays a whole number, and the total $\sum a_i$ is unchanged (we moved one unit from a to b). Watch the sum of squares. Replacing a, b by $a-1, b+1$ changes it by

$$(a-1)^2 + (b+1)^2 - a^2 - b^2 = -2a + 1 + 2b + 1 = 2(b - a + 1).$$

Since $a \geq b+2$, we have $b - a + 1 \leq -1$, so $\sum a_i^2$ *drops by at least* 2 on every move.

Why it must stop. $\sum a_i^2$ is a non-negative integer, and a non-negative integer cannot decrease forever — it would eventually fall below 0. So only finitely many moves are possible, and the process halts. It can stop only when no two values differ by 2 or more, i.e. when all the numbers are within 1 of one another. (**Contrast the real-valued "replace by the average" process, which can shrink forever without ever terminating — the integer steps are what force a stop.**) ■

Problem 14. An equilateral triangle of side n is cut into n^2 unit triangles. Prove it can never be tiled by rhombi (each rhombus being one up-triangle joined to one down-triangle).

SOLUTION. Two-color up against down, then count. Color every upward-pointing unit triangle one color and every downward-pointing one another. The defining feature of the rhombus is that it is exactly one up-triangle glued to one adjacent down-triangle, so *every* rhombus — in any of its orientations — covers precisely one up and one down.

Now count the two types in a side-n triangle. Reading the rows from the apex down, the upward triangles number $1+2+\cdots+n = \frac{n(n+1)}{2}$, while the downward ones number $0+1+\cdots+(n-1) = \frac{n(n-1)}{2}$. Their difference is

$$\frac{n(n+1)}{2} - \frac{n(n-1)}{2} = n,$$

so there are always exactly n more upward triangles than downward. A rhombus tiling would consume the two types in equal numbers, which is impossible when they differ by $n \geq 1$. Hence no rhombus tiling exists, for any n. ■

Count Something Else

Full worked solutions. Each one builds an explicit matching, then checks it runs perfectly both ways.

Launch

Problem 1. How many ways can 5 identical candies be given to 3 children (some may get none)?

SOLUTION. Stars and bars. Represent the 5 candies as stars and use 2 bars to split them among 3 children. Each arrangement of 5 stars and 2 bars in a row encodes one distribution, and choosing the 2 bar-positions among the $5 + 2 = 7$ slots gives

$$\binom{7}{2} = 21.$$

Problem 2. Show a set of n elements has equally many k-subsets and $(n-k)$-subsets, via an explicit matching.

SOLUTION. Match each subset with its complement. Send each k-element subset A to its complement $\overline{A} = \{1, \dots, n\} \setminus A$, which has $n - k$ elements. This map is its own inverse (the complement of the complement is the original set), so it pairs k-subsets with $(n-k)$-subsets perfectly — a bijection. Hence $\binom{n}{k} = \binom{n}{n-k}$. ■

Core

Problem 3. How many monotone lattice paths go from $(0,0)$ to (m,n) using unit right/up steps?

SOLUTION. Encode the path as a string. Every such path is a sequence of m right-steps (R) and n up-steps (U), in some order; conversely every arrangement of m R's and n U's traces exactly one path. This is a bijection between paths and strings of length $m+n$ with m R's. Counting the strings by choosing which positions hold the R's:

$$\binom{m+n}{m}.$$

$\binom{m+n}{m}$

Problem 4. Prove that even-sized and odd-sized subsets of $\{1,\dots,n\}$ are equal in number, via the toggle bijection.

SOLUTION. Toggle element #1. Define a map on subsets: if the subset contains element 1, remove it; if not, add it. This changes the subset's size by exactly 1, so it turns every even-sized subset into an odd-sized one and vice versa. Applying it twice returns the original subset, so it is its own inverse — a bijection. It therefore pairs the even-sized subsets one-to-one with the odd-sized subsets, forcing the two collections to have equal size (each 2^{n-1} for $n \geq 1$). ■

Problem 5. How many ways can n be written as an ordered sum of k *positive* parts?

SOLUTION. Reduce to stars and bars. A composition into k positive parts requires each part ≥ 1. Give every part one unit up front; this uses k units and leaves $n-k$ units to distribute among the k parts with *no* positivity restriction. By stars and bars, the number of ways to split $n-k$ among k parts is $\binom{(n-k)+(k-1)}{k-1} = \binom{n-1}{k-1}$. **So there are $\binom{n-1}{k-1}$ compositions of n into k positive parts.**

$\binom{n-1}{k-1}$

Stretch

Problem 6. Show the number of ways to nest n pairs of parentheses equals C_n, via a bijection to balanced up/down paths.

SOLUTION. Match each parenthesis with a step. Read a string of n left and n right parentheses left to right; send each "(" to an up-step $(+1)$ and each ")" to a down-step (-1). A string of parentheses is *validly nested* exactly when, at every prefix, the number of ")" never exceeds the number of "(" —

equivalently, the path's running height never drops below 0. This is a bijection between valid nestings and balanced up/down paths of length $2n$ that stay ≥ 0 and end at height 0 (Dyck paths). The number of Dyck paths is the Catalan number C_n, so the nestings number C_n as well. ■

Problem 7. Prove $\sum_{k=0}^{n} \binom{n}{k}^2 = \binom{2n}{n}$ by counting lattice paths through a midpoint two ways.

SOLUTION. Count monotone paths from $(0,0)$ to (n,n). By Problem 3, there are $\binom{2n}{n}$ such paths.

Now count them by where they cross the "anti-diagonal" midline $x + y = n$. A path from $(0,0)$ to (n,n) hits the midline at exactly one lattice point $(k, n-k)$ for some $0 \leq k \leq n$. The number of paths from $(0,0)$ to $(k, n-k)$ is $\binom{n}{k}$, and the number from $(k, n-k)$ to (n,n) is $\binom{n}{n-k}$ (it needs $n-k$ more rights and k more ups). Since $\binom{n}{n-k} = \binom{n}{k}$, the paths through that midpoint number $\binom{n}{k}^2$. Summing over all crossing points,

$$\binom{2n}{n} = \sum_{k=0}^{n} \binom{n}{k}\binom{n}{n-k} = \sum_{k=0}^{n} \binom{n}{k}^2.$$

■

The Flash

Problem 8. Prove $\sum_{k=0}^{n} \binom{n}{k}^2 = \binom{2n}{n}$ by a committee argument.

SOLUTION. One committee, counted two ways. Form a club of n women and n men and choose a committee of exactly n people from the $2n$ total. Counted directly, there are $\binom{2n}{n}$ such committees. Counted by how many women it contains: if the committee has k women, it must take the other $n-k$ members from the men, giving $\binom{n}{k}\binom{n}{n-k}$ ways. Since $\binom{n}{n-k} = \binom{n}{k}$, that is $\binom{n}{k}^2$. Summing over k,

$$\sum_{k=0}^{n} \binom{n}{k}^2 = \binom{2n}{n}.$$

(Check at $n = 2$: $1 + 4 + 1 = 6 = \binom{4}{2}$.) This is the central case of Vandermonde's identity. ■

$\binom{2n}{n}$

Problem 9. How many non-negative integer solutions does $a + b + c + d = 10$ have?

SOLUTION. Stars and bars with 10 stars and 3 bars (for 4 variables):

$$\binom{10+3}{3} = \binom{13}{3} = 286.$$

286

Problem 10. In how many ways can 6 people around a round table be split into pairs of adjacent neighbors?

SOLUTION. Reframe as covering a cycle by dominoes. Seat the 6 people at positions $1, \dots, 6$ around a circle. Pairing adjacent neighbors means tiling the 6-cycle by "dominoes" (adjacent pairs). A cycle of even length $2m$ has exactly 2 perfect matchings into adjacent pairs (the two ways of choosing alternate edges). For 6 people ($m = 3$): the pairs are either $\{1,2\},\{3,4\},\{5,6\}$ or $\{2,3\},\{4,5\},\{6,1\}$. **There are 2 such pairings.**

2

Proof Lab

Problem 11. Prove the reflection principle and derive $C_n = \binom{2n}{n} - \binom{2n}{n+1}$.

SOLUTION. Reflect the bad paths. Count up/down paths of $2n$ steps starting at height 0 that end at height 0. Total paths: choose which n of the $2n$ steps go up, $\binom{2n}{n}$. Call a path *bad* if it ever dips to height -1.

Given a bad path, locate the *first* time it touches -1, and reflect every step *after* that moment across the line $y = -1$ (swap up↔down). The reflected path starts at 0 but now ends at height -2 instead of 0. This reflection is reversible (reflect back from the first touch of -1), so it is a bijection between bad paths and *all* paths from 0 to -2. The number of paths from 0 to -2 in $2n$ steps is $\binom{2n}{n+1}$ (one more down-step than up). Therefore

$$C_n = (\text{all paths}) - (\text{bad paths}) = \binom{2n}{n} - \binom{2n}{n+1} = \frac{1}{n+1}\binom{2n}{n}.$$

■

Problem 12. Give a bijection between binary trees with n internal nodes and triangulations of a convex $(n+2)$-gon, concluding both count C_n.

SOLUTION. Split at a fixed edge. Fix one edge of the convex $(n+2)$-gon as the "base." In any triangulation, the base edge belongs to exactly one triangle, whose apex is some other vertex. That triangle splits the polygon into a smaller polygon on its left and one on its right, each triangulated. Map this to a binary tree: the chosen triangle is the root, and the left and right sub-polygons become the left and right subtrees, recursively.

This correspondence is reversible — given a binary tree, rebuild the triangulation by assigning the root triangle to the base edge and recursing — so it is a bijection. Both objects therefore satisfy the same recurrence and initial conditions as the Catalan numbers, and both are counted by C_n. ■

Problem 13. Prove that partitions of n into distinct parts are equinumerous with partitions of n into odd parts.

SOLUTION. Pull out the powers of two. Build a map from distinct-part partitions to odd-part partitions. Every positive integer factors uniquely as (odd number) $\times 2^a$. Take a partition of n into *distinct* parts; write each part that way, then replace each part $2^a \cdot m$ (with m odd) by 2^a copies of the odd number m. The total being summed is unchanged, so the result is a partition of n into *odd* parts.

To see this is a bijection, reverse it: given a partition into odd parts, group the equal parts. If an odd value m appears c times, write the multiplicity c in binary, $c = 2^{a_1} + 2^{a_2} + \cdots$ with distinct exponents, and produce the distinct parts $2^{a_1}m$, $2^{a_2}m, \ldots$. Because each exponent in a binary expansion is distinct *and* the odd "cores" m differ between values, all the parts produced are distinct, and the construction inverts the forward map exactly. Hence the two counts are equal. ■

Equal, via the odd $\times 2^a$ correspondence

Problem 14. Bijection between balanced parentheses and non-crossing handshakes; both are C_n.

SOLUTION. Nesting is non-crossing. Take a balanced string of n pairs of parentheses and seat its $2n$ symbols at $2n$ points around a circle in order. Each "(" has a unique matching ")" (the one that closes it); draw a chord between their seats. *Two chords never cross:* matched pairs in a balanced string are either nested (one interval inside the other) or disjoint (side by side), and in both cases the chords avoid each other — a crossing would require the pattern ([ )], which is not properly nested. Conversely, from a non-crossing perfect matching, walk around the circle writing "(" when a chord's first endpoint is met and ")" at its second; non-crossing guarantees the string is balanced. The two constructions are inverse, so balanced strings and non-crossing handshakes are equinumerous — and since balanced strings of n pairs are the Dyck paths counted earlier, both equal C_n. ■

Both are C_n

Problem 15. Prove there are n^{n-2} labeled trees, via Prüfer sequences.

SOLUTION. Encode a tree as a length-$(n-2)$ string of labels. Given a labeled tree on $n \geq 2$ vertices, repeat $n-2$ times: find the leaf of smallest label, record the label of its unique neighbor, and delete that leaf. This yields a sequence in $\{1, \ldots, n\}^{n-2}$.

To invert it, note a vertex's degree equals (its number of appearances in the sequence) $+1$; so the vertices *not* yet appearing are exactly the current leaves. Repeatedly take the smallest such leaf, join it to the first remaining sequence entry, and cross that entry off; two vertices remain at the end and are joined by the last edge. This reconstructs the tree uniquely, so encoding is a bijection between trees and sequences. There are n^{n-2} sequences, hence n^{n-2} labeled trees.

Example ($n = 4$). The path $1-3-2-4$ has leaves 1 and 4; smallest leaf 1 has neighbor 3 (record 3, delete 1), then leaves are 3, 4, smallest leaf 3 has neighbor 2 (record 2). Prüfer sequence $(3, 2)$ — length $4 - 2 = 2$, as expected, and $4^2 = 16$ trees in all. ■

n^{n-2}

Problem 16. Prove Fermat's little theorem $a^p \equiv a \pmod{p}$ by counting necklaces.

SOLUTION. Rotations sort strings into orbits of size 1 or p. Consider all a^p strings of p beads, each bead one of a colors. Rotation by one position acts on these strings; group each string with its rotations. A string is fixed by a nontrivial rotation only if it is constant (all beads equal): if a shift by d with $1 \leq d \leq p-1$ fixes a string, then since $\gcd(d, p) = 1$ every shift fixes it, forcing all beads equal. There are exactly a such monochromatic strings, each its own orbit of size 1. Every other string has all p rotations distinct, so its orbit has size exactly p. Counting the a^p strings by orbits,

$$a^p = a + p \cdot (\text{number of size-}p\text{ orbits}),$$

so $p \mid a^p - a$, i.e. $a^p \equiv a \pmod{p}$. ■

$a^p \equiv a \pmod{p}$

Solve It Smaller

Full worked solutions. Each one finds how a size-n problem reduces to smaller copies, then builds up — and proves the pattern really holds.

Launch

Problem 1. Climbing n stairs in steps of 1 or 2: how many ways? Set up the recurrence.

SOLUTION. Split on the last step. Your final move onto stair n is either a single step (from stair $n-1$) or a double step (from stair $n-2$). These cases are disjoint and cover everything, so if S_n counts the ways,

$$S_n = S_{n-1} + S_{n-2}, \qquad S_1 = 1,\ S_2 = 2.$$

This is the Fibonacci recurrence; the sequence runs $1, 2, 3, 5, 8, \ldots$, so $S_n = F_{n+1}$. **The answer is a Fibonacci number.**

$$S_n = F_{n+1}$$

Problem 2. $a_1 = 3$, $a_n = 2a_{n-1}$. Find a_6 and a closed form.

SOLUTION. Each term doubles the previous, so $a_n = 3 \cdot 2^{n-1}$. Then $a_6 = 3 \cdot 2^5 = 96$.

$$a_6 = 96,\ a_n = 3 \cdot 2^{n-1}$$

Problem 3. Prove by induction that $1 + 3 + 5 + \cdots + (2n-1) = n^2$ for all $n \geq 1$.

SOLUTION. Base case. For $n = 1$ the left side is the single term 1 and the right side is $1^2 = 1$. They agree. ✓

Inductive step. Assume the identity at k: $1 + 3 + \cdots + (2k-1) = k^2$. Add the next odd number, $2k+1$, to both sides:

$$\underbrace{1 + 3 + \cdots + (2k-1)}_{=k^2} + (2k+1) = k^2 + 2k + 1 = (k+1)^2.$$

The left side is now the sum of the first $k+1$ odd numbers, and the right side is $(k+1)^2$ — exactly the claim at $k+1$. By induction it holds for all $n \geq 1$. ■

Core

Problem 4. How many length-n binary strings have no two adjacent 1s?

SOLUTION. Split on the last bit. Let f_n count the valid strings. If the last bit is 0, the first $n-1$ bits form any valid string of length $n-1$: that's f_{n-1}. If the last bit is 1, the bit before it must be 0 (no adjacent 1s), and the first $n-2$ bits form any valid string: that's f_{n-2}. The cases are disjoint and exhaustive, so

$$f_n = f_{n-1} + f_{n-2}, \qquad f_1 = 2, f_2 = 3.$$

Again Fibonacci: $f_n = F_{n+2}$. ■

Problem 5. Solve $a_n = 2a_{n-1} + 3$, $a_0 = 0$.

SOLUTION. Compute and spot the pattern. The terms are $0, 3, 9, 21, 45, 93, \ldots$, each one short of a multiple of 3: indeed $0 = 3(2^0 - 1), 3 = 3(2^1 - 1), 9 = 3(2^2 - 1), 21 = 3(2^3 - 1)$. Conjecture $a_n = 3(2^n - 1)$.

Verify. It satisfies the recurrence: $2 \cdot 3(2^{n-1} - 1) + 3 = 3(2^n - 2) + 3 = 3(2^n - 1)$, and $a_0 = 3(1-1) = 0$. So $a_n = 3(2^n - 1)$.

$$a_n = 3(2^n - 1)$$

Problem 6. Count the regions inside a circle cut by n chords in general position. Build the recurrence.

SOLUTION. Add one chord at a time. Suppose R_{n-1} regions exist; draw the n-th chord. It crosses each of the $n-1$ existing chords once (general position), at $n-1$ interior points, which split the new chord into n arcs. Each arc cuts an existing region into two, adding exactly one region per arc — so the n-th chord adds n regions:

$$R_n = R_{n-1} + n, \qquad R_0 = 1.$$

Unrolling, $R_n = 1 + (1 + 2 + \cdots + n) = 1 + \binom{n+1}{2}$. (Check $n = 5$: $1 + 15 = 16$.) ■

Problem 7. Prove by induction that $n < 2^n$ for every integer $n \geq 1$.

SOLUTION. Base case. For $n = 1$: $1 < 2^1 = 2$. ✓

Inductive step. Assume $k < 2^k$ for some $k \geq 1$. Then

$$k + 1 \;\leq\; 2^k + 1 \;\leq\; 2^k + 2^k \;=\; 2^{k+1},$$

where the first inequality uses the hypothesis and the second uses $1 \leq 2^k$ (true for $k \geq 0$). In fact the inequality is strict, since $k + 1 \leq 2^k + 1 < 2^k + 2^k$ whenever $2^k > 1$, i.e. $k \geq 1$. Thus $k + 1 < 2^{k+1}$, and by induction $n < 2^n$ for all $n \geq 1$. ■

Problem 8. Solve $a_n = 5a_{n-1} - 6a_{n-2}$, $a_0 = 0$, $a_1 = 1$, via the characteristic equation.

SOLUTION. Form the characteristic equation. Guessing $a_n = r^n$ gives $r^2 = 5r - 6$, i.e. $r^2 - 5r + 6 = (r-2)(r-3) = 0$, with distinct roots $r = 2$ and $r = 3$. So the general solution is

$$a_n = \alpha\, 2^n + \beta\, 3^n.$$

Fit the seeds. From $a_0 = 0$: $\alpha + \beta = 0$, so $\beta = -\alpha$. From $a_1 = 1$: $2\alpha + 3\beta = 2\alpha - 3\alpha = -\alpha = 1$, so $\alpha = -1$ and $\beta = 1$. Therefore

$$a_n = 3^n - 2^n.$$

(Check: $a_2 = 9 - 4 = 5$, and the recurrence gives $5 \cdot 1 - 6 \cdot 0 = 5$. ✓) ■

$a_n = 3^n - 2^n$

Stretch

Problem 9. Prove the derangement recurrence $D_n = (n-1)(D_{n-1} + D_{n-2})$ by splitting on where element 1 lands.

SOLUTION. Split on the image of element 1. In a derangement of $\{1, \dots, n\}$, element 1 maps to some position $j \neq 1$, and there are $n - 1$ choices for j. Fix one such j; split into two cases by what maps to position 1:

Case A: element j maps back to 1 (so 1 and j swap). The remaining $n - 2$ elements must derange among themselves: $D_{n\ 2}$ ways.

Case B: element j does not *map to* 1. Some other element m (with $m \neq 1$, and $m \neq j$ since $j \not\to 1$) is the one sent to position 1. Splice 1 out: reroute m straight to j — the slot 1 was heading for — and delete element 1 and position 1 entirely. What remains is a permutation of the $n - 1$ symbols $\{2, \dots, n\}$, and it is a *derangement*: the new edge $m \to j$ is not a fixed point (since $m \neq j$), and every other element still avoids its own place. The splice is reversible — from any derangement of $\{2, \dots, n\}$, locate the element m sitting where j is used, then insert 1 by setting $m \to 1$ and $1 \to j$ — so it is a bijection. Hence exactly D_{n-1} derangements arise this way.

The two cases are disjoint and exhaustive, and there were $n-1$ choices of j, so $D_n = (n-1)(D_{n-1} + D_{n-2})$. ■

Problem 10. Count triangulations of a convex $(n+2)$-gon; set up the Catalan recurrence.

SOLUTION. Split on the triangle attached to a fixed edge. Fix one edge of the $(n+2)$-gon as the base. In any triangulation, the base lies in exactly one triangle, whose third vertex divides the polygon into a sub-polygon to the left (with $k + 2$ vertices, say) and one to the right (with the complementary number). Each sub-polygon is independently triangulated. If T_m denotes the number of triangulations of an $(m+2)$-gon, summing over the position of the apex gives

$$T_n = \sum_{i=0}^{n-1} T_i T_{n-1-i},$$

the Catalan convolution, so $T_n = C_n$. ■

Problem 11. Pell numbers $P_n = 2P_{n-1} + P_{n-2}$, $P_0 = 0$, $P_1 = 1$: find a Binet-style closed form.

SOLUTION. Form the characteristic equation. Guessing $P_n = r^n$ gives $r^2 = 2r + 1$, i.e. $r^2 - 2r - 1 = 0$, with roots

$$r = \frac{2 \pm \sqrt{4+4}}{2} = 1 \pm \sqrt{2}.$$

Write $\delta = 1 + \sqrt{2}$ (the *silver ratio*) and $\bar{\delta} = 1 - \sqrt{2}$. The general solution is $P_n = \alpha\,\delta^n + \beta\,\bar{\delta}^n$.

Fit the seeds. From $P_0 = 0$: $\alpha + \beta = 0$, so $\beta = -\alpha$. From $P_1 = 1$: $\alpha\delta + \beta\bar{\delta} = \alpha(\delta - \bar{\delta}) = \alpha\,(2\sqrt{2}) = 1$, so $\alpha = \dfrac{1}{2\sqrt{2}}$. Therefore

$$P_n = \frac{(1+\sqrt{2})^n - (1-\sqrt{2})^n}{2\sqrt{2}}.$$

(Check: $P_2 = \dfrac{(3+2\sqrt{2}) - (3-2\sqrt{2})}{2\sqrt{2}} = \dfrac{4\sqrt{2}}{2\sqrt{2}} = 2$, matching $2 \cdot 1 + 0$. ✓) Just as with Binet's formula, the irrational pieces conspire to give an integer at every step. ■

$$P_n = \frac{(1+\sqrt{2})^n - (1-\sqrt{2})^n}{2\sqrt{2}}$$

Problem 12. Roots r, s of $x^2 - x - 1 = 0$: show $p_k = r^k + s^k$ satisfies $p_k = p_{k-1} + p_{k-2}$, and compute $r^5 + s^5$.

SOLUTION. The power sums inherit the recurrence. Each root satisfies its own equation: $r^2 = r + 1$, so multiplying by r^{k-2} gives $r^k = r^{k-1} + r^{k-2}$, and likewise $s^k = s^{k-1} + s^{k-2}$. Adding the two,

$$p_k = r^k + s^k = (r^{k-1} + s^{k-1}) + (r^{k-2} + s^{k-2}) = p_{k-1} + p_{k-2}.$$

Seed and iterate. $p_0 = r^0 + s^0 = 2$ and $p_1 = r + s = 1$ (the coefficient of x). So the sequence runs $2, 1, 3, 4, 7, 11, \ldots$ — the Lucas numbers — and $p_5 = 11$. No $\sqrt{5}$ ever appeared. ■

$$r^5 + s^5 = 11$$

The Flash

Problem 13. How many tilings does a 2×10 strip have?

SOLUTION. The tiling count satisfies $T_n = T_{n-1} + T_{n-2}$ with $T_1 = 1, T_2 = 2$, i.e. $T_n = F_{n+1}$. So $T_{10} = F_{11} = 89$.

89

Problem 14. Tower of Hanoi with $n = 8$ disks: how many moves?

SOLUTION. The minimal move count is $H_n = 2^n - 1$. For $n = 8$, $H_8 = 2^8 - 1 = 255$.

255

Problem 15. Roots r, s of $x^2 - 3x + 1 = 0$: find $r^4 + s^4$ via the power-sum recurrence.

SOLUTION. From $x^2 = 3x - 1$, the power sums obey $p_k = 3p_{k-1} - p_{k-2}$, with $p_0 = 2$ and $p_1 = r + s = 3$. Iterate: $p_2 = 3 \cdot 3 - 2 = 7$, then $p_3 = 3 \cdot 7 - 3 = 18$, then $p_4 = 3 \cdot 18 - 7 = 47$. So $r^4 + s^4 = 47$.

47

Proof Lab

Problem 16. Prove the $2 \times n$ domino tiling count equals F_{n+1}, with the last-column split shown exhaustive and disjoint.

SOLUTION. Classify by how the last column is covered. Let T_n be the number of tilings of a $2 \times n$ strip by 1×2 dominoes. Consider the rightmost column. It is covered in exactly one of two mutually exclusive ways:

(i) *A single vertical domino* fills the last column. Removing it leaves a $2 \times (n-1)$ strip, tiled in T_{n-1} ways.

(ii) *Two horizontal dominoes* cover the last column's two cells; their left halves occupy the second-to-last column. These two horizontals together fill the last *two* columns' relevant cells, leaving a $2 \times (n-2)$ strip, tiled in T_{n-2} ways.

No tiling can do both (the top-right cell is covered by exactly one domino, vertical or horizontal), and every tiling does one, so the split is disjoint and exhaustive:

$$T_n = T_{n-1} + T_{n-2}, \qquad T_1 = 1,\ T_2 = 2.$$

This is the Fibonacci recurrence with the indicated initial values, so $T_n = F_{n+1}$. ■

Problem 17. Prove $\sum_{i=0}^{n} C_i C_{n-i} = C_{n+1}$ counts parenthesizations of $n+2$ factors, and explain the binary-tree link.

SOLUTION. Split on the outermost multiplication. A full parenthesization of $a_0 a_1 \cdots a_{n+1}$ ($n+2$ factors) has exactly one *last* multiplication, joining a parenthesized left block (the first $i+1$ factors) with a parenthesized right block (the remaining $n+1-i$ factors), for some split point i. The left block, with $i+1$ factors, can be parenthesized in C_i ways; the right block, with $n+1-i$ factors, in C_{n-i} ways. These choices are independent, and summing over the split point $i = 0, \dots, n$,

$$C_{n+1} = \sum_{i=0}^{n} C_i\, C_{n-i}.$$

Binary trees. The same recurrence governs binary trees because a binary tree with a root splits into an independent left subtree and right subtree — exactly the "left block, right block" decomposition. The last multiplication corresponds to the root; the two blocks to the two subtrees. Hence parenthesizations, binary trees, and triangulations all obey the Catalan recurrence. ■

Problem 18. Prove by strong induction that every amount of postage ≥ 12 cents can be made with 4- and 5-cent stamps.

SOLUTION. Four base cases. The step below will reach back four units, so we verify the four smallest amounts directly:

$$12 = 4+4+4, \quad 13 = 4+4+5, \quad 14 = 4+5+5, \quad 15 = 5+5+5. \checkmark$$

Strong inductive step. Let $n \geq 16$, and assume every amount from 12 up to $n-1$ is makeable. Since $n \geq 16$, we have $12 \leq n-4 \leq n-1$, so by the strong hypothesis $n-4$ cents can be formed from 4- and 5-cent stamps. Adding one more 4-cent stamp produces exactly n cents. By strong induction, every amount ≥ 12 is makeable. ■

Remark. Four base cases are essential, not decorative: the step only converts "$n-4$ works" into "n works," so without the block 12–15 verified by hand, the chain would have nothing to stand on. This is the trapdoor from the chapter — a step that reaches back four steps needs four anchors.

Problem 19. Roots r, s of $x^2 - x - 1 = 0$: prove $p_n = r^n + s^n$ is an integer for all $n \geq 0$; identify the sequence.

SOLUTION. An integer-seeded recurrence. Each root gives $r^n = r^{n-1} + r^{n-2}$ (multiply $r^2 = r+1$ by r^{n-2}), and adding the twin for s shows the power sums satisfy

$$p_n = p_{n-1} + p_{n-2}, \qquad p_0 = 2, \quad p_1 = 1.$$

Induction. The seeds $p_0 = 2$ and $p_1 = 1$ are integers. If p_{n-1} and p_{n-2} are integers, then $p_n = p_{n-1} + p_{n-2}$ is a sum of integers, hence an integer; by strong induction *every* p_n is an integer — even though r^n and s^n separately involve $\sqrt{5}$. The sequence $2, 1, 3, 4, 7, 11, 18, \dots$ is exactly the *Lucas numbers* L_n. ■

Problem 20. Prove the Josephus halving recurrence and closed form; find the survivor for $n =$ 1000.

SOLUTION. One lap halves the circle; then relabel. Number the seats 1 to the circle size and eliminate every second seat starting from 2.

Even size $2n$. The first lap eliminates $2, 4, \dots, 2n$, leaving the n odd seats $1, 3, \dots, 2n-1$, with the count about to resume at seat 1. This is precisely a fresh game of size n: seat j of the small game is the j-th odd number, $2j-1$, here. So if the small game's survivor is $J(n)$, the big game's is

$$J(2n) = 2J(n) - 1.$$

Odd size $2n+1$. The first lap eliminates $2, 4, \dots, 2n$; then seat $2n+1$ is counted safe and seat 1 is eliminated. The n survivors $3, 5, \dots, 2n+1$ resume the count at seat 3, and small-game seat j maps to $2j+1$:

$$J(2n+1) = 2J(n) + 1, \qquad J(1) = 1.$$

Closed form. Write $n = 2^m + \ell$ with $0 \le \ell < 2^m$. Peeling seats off with the two rules is, in binary, removing the leading 1 (subtracting 2^m, leaving ℓ) and re-appending it at the bottom (doubling and adding 1); the net effect is $J(n) = 2\ell + 1$. Formally, induct on n: if the formula holds below n, then writing $n = 2q$ or $2q+1$ and applying the matching rule reproduces $2\ell + 1$. For $n = 1000 = 512 + 488$ we have $\ell = 488$, so $J(1000) = 2(488) + 1 = 977$. ■

$J(1000) = 977$

Problem 21. Josephus, every third: justify the recurrence and compute the survivor for $n = 7$.

SOLUTION. Track the survivor's seat as the circle shrinks by one. Number the seats $0, 1, \dots, n-1$ and eliminate every third. The very first elimination removes seat 2 (the third seat counted from 0); the count then restarts at seat 3 on a circle of $n-1$ people. Relabel that smaller circle so its count starts at 0: every surviving seat is shifted down by 3 (equivalently, the small-circle answer must be shifted *up* by 3 to name the original seat), and because the seats wrap around, the shift is taken modulo the current size. Hence

$$J(n) = (J(n-1) + 3) \bmod n, \qquad J(1) = 0.$$

Iterating: $J(1) = 0$, $J(2) = 3 \bmod 2 = 1$, $J(3) = 4 \bmod 3 = 1$, $J(4) = 4 \bmod 4 = 0$, $J(5) = 3 \bmod 5 = 3$, $J(6) = 6 \bmod 6 = 0$, $J(7) = 3 \bmod 7 = 3$. Since seats were numbered from 0, the survivor counted from 1 is seat 4. ■

Seat 4

The Great Cancellation

Full worked solutions. Each one writes the term as a difference $f(k) - f(k+d)$ and reads off the surviving endpoints.

Launch

Problem 1. Evaluate $\sum_{k=1}^{n}(k^2 - (k-1)^2)$ by telescoping.

SOLUTION. Each term is already a difference $f(k) - f(k-1)$ with $f(k) = k^2$. Writing it out,

$$(1^2 - 0^2) + (2^2 - 1^2) + \cdots + (n^2 - (n-1)^2),$$

every interior square is created once and cancelled once; only -0^2 and $+n^2$ survive. The sum is $n^2 - 0 = n^2$. (As a check, the summand simplifies to $2k - 1$, and $\sum_{k=1}^{n}(2k-1) = n^2$ is the classic odd-number identity.)

n^2

Problem 2. Find $\sum_{k=1}^{20} \frac{1}{k(k+1)}$.

SOLUTION. Using $\frac{1}{k(k+1)} = \frac{1}{k} - \frac{1}{k+1}$, the sum telescopes to $1 - \frac{1}{21} = \frac{20}{21}$.

$\frac{20}{21}$

Core

Problem 3. Evaluate $\sum_{k=1}^{n} \frac{1}{k(k+1)(k+2)}$ by partial fractions.

SOLUTION. Write the term as a difference of adjacent products. Verify the hint:

$$\frac{1}{2}\left(\frac{1}{k(k+1)}-\frac{1}{(k+1)(k+2)}\right)=\frac{1}{2}\cdot\frac{(k+2)-k}{k(k+1)(k+2)}=\frac{1}{k(k+1)(k+2)}.$$

So with $f(k)=\frac{1}{k(k+1)}$, the summand is $\frac{1}{2}(f(k)-f(k+1))$, and the sum telescopes:

$$\sum_{k=1}^{n}\frac{1}{k(k+1)(k+2)}=\frac{1}{2}\left(\frac{1}{1\cdot 2}-\frac{1}{(n+1)(n+2)}\right)=\frac{1}{2}\left(\frac{1}{2}-\frac{1}{(n+1)(n+2)}\right).$$

(Check $n=5$: $\frac{1}{2}(\frac{1}{2}-\frac{1}{42})=\frac{5}{21}$.) ■

Problem 4. Find $\sum_{k=1}^{n}\frac{k}{(k+1)!}$.

SOLUTION. Spot the factorial difference. Write $\frac{k}{(k+1)!}=\frac{(k+1)-1}{(k+1)!}=\frac{1}{k!}-\frac{1}{(k+1)!}$. With $f(k)=\frac{1}{k!}$, the summand is $f(k)-f(k+1)$, so the sum telescopes:

$$\sum_{k=1}^{n}\frac{k}{(k+1)!}=\frac{1}{1!}-\frac{1}{(n+1)!}=1-\frac{1}{(n+1)!}.$$

$1-\frac{1}{(n+1)!}$

Problem 5. Evaluate $\prod_{k=2}^{n}\frac{k^3-1}{k^3+1}$.

SOLUTION. Factor sum and difference of cubes. Use $k^3-1=(k-1)(k^2+k+1)$ and $k^3+1=(k+1)(k^2-k+1)$. Note also that $k^2-k+1=(k-1)^2+(k-1)+1$, i.e. the "k^2+k+1" pattern shifted down by one. So

$$\frac{k^3-1}{k^3+1}=\frac{(k-1)}{(k+1)}\cdot\frac{k^2+k+1}{(k-1)^2+(k-1)+1}.$$

The first factors telescope as a product: $\prod_{k=2}^{n}\frac{k-1}{k+1}=\frac{1\cdot 2}{n(n+1)}$. The second factors telescope too, leaving $\frac{n^2+n+1}{2^2-2+1}=\frac{n^2+n+1}{3}$. Multiplying,

$$\prod_{k=2}^{n}\frac{k^3-1}{k^3+1}=\frac{2}{n(n+1)}\cdot\frac{n^2+n+1}{3}=\frac{2(n^2+n+1)}{3n(n+1)}.$$

(Check $n=5$: $\frac{2\cdot 31}{3\cdot 30}=\frac{62}{90}=\frac{31}{45}$.) ■

Stretch

Problem 6. Derive $\sum_{k=1}^{n}k^2=\frac{n(n+1)(2n+1)}{6}$ by telescoping $k^3-(k-1)^3$.

SOLUTION. Telescope a cube difference, then solve. The sum $\sum_{k=1}^{n}(k^3-(k-1)^3)$ telescopes to n^3. But expanding the summand,

$$k^3-(k-1)^3=3k^2-3k+1,$$

so $\sum_{k=1}^{n}(3k^2-3k+1)=n^3$. That is $3\sum k^2-3\sum k+n=n^3$. Using $\sum k=\frac{n(n+1)}{2}$,

$$3\sum k^2=n^3+\frac{3n(n+1)}{2}-n=n^3-n+\frac{3n(n+1)}{2}.$$

Simplifying gives $\sum k^2=\frac{n(n+1)(2n+1)}{6}$. ■

Problem 7. Evaluate $\sum_{k=1}^{n}\frac{k}{2^k}$ by finding g with $\frac{k}{2^k}=g(k)-g(k+1)$.

SOLUTION. Guess $g(k)=\frac{ak+b}{2^{k-1}}$ and match. Matching $g(k)-g(k+1)$ against $\frac{k}{2^k}$ pins down $a=b=1$, giving $g(k)=\frac{k+1}{2^{k-1}}$. Then

$$g(k)-g(k+1)=\frac{k+1}{2^{k-1}}-\frac{k+2}{2^k}=\frac{2(k+1)-(k+2)}{2^k}=\frac{k}{2^k}.$$

So the summand is $g(k)-g(k+1)$ and the sum telescopes:

$$\sum_{k=1}^{n}\frac{k}{2^k}=g(1)-g(n+1)=\frac{2}{2^0}-\frac{n+2}{2^n}=2-\frac{n+2}{2^n}.$$

(Check $n=4$: $2-\frac{6}{16}=\frac{13}{8}$.) ■

Problem 7b. Derive $\sum_{k=1}^{n}k^3=\left(\frac{n(n+1)}{2}\right)^2$ by telescoping $(k+1)^4-k^4$.

SOLUTION. Telescope one power higher, then substitute the lower sums. The sum $\sum_{k=1}^{n}((k+1)^4-k^4)$ telescopes to $(n+1)^4-1$. Expanding the summand,

$$(k+1)^4-k^4=4k^3+6k^2+4k+1,$$

so $4\sum k^3+6\sum k^2+4\sum k+n=(n+1)^4-1$. Substitute the known sums $\sum k^2=\frac{n(n+1)(2n+1)}{6}$ and $\sum k=\frac{n(n+1)}{2}$ and solve:

$$4\sum k^3=(n+1)^4-1-n-6\cdot\tfrac{n(n+1)(2n+1)}{6}-4\cdot\tfrac{n(n+1)}{2},$$

which simplifies to $\sum k^3=\frac{n^2(n+1)^2}{4}=\left(\frac{n(n+1)}{2}\right)^2$ — the square of the sum of the first n integers. (Check $n=4$: $\left(\frac{4\cdot5}{2}\right)^2=100=1+8+27+64$.) ■

The Flash

Problem 8. Compute $\sum_{k=1}^{99}\frac{1}{\sqrt{k}+\sqrt{k+1}}$.

SOLUTION. Rationalize each term by multiplying by $\frac{\sqrt{k+1}-\sqrt{k}}{\sqrt{k+1}-\sqrt{k}}$:

$$\frac{1}{\sqrt{k}+\sqrt{k+1}} = \frac{\sqrt{k+1}-\sqrt{k}}{(k+1)-k} = \sqrt{k+1}-\sqrt{k}.$$

This telescopes: $\sum_{k=1}^{99}(\sqrt{k+1}-\sqrt{k}) = \sqrt{100}-\sqrt{1} = 10-1 = 9$.

9

Problem 9. Compute $\sum_{k=1}^{n} k \cdot k!$.

SOLUTION. Since $k \cdot k! = (k+1)! - k!$, the sum telescopes to $(n+1)! - 1! = (n+1)! - 1$.

$(n+1)! - 1$

Proof Lab

Problem 10. Prove $\sum_{k=1}^{n} (f(k) - f(k+1)) = f(1) - f(n+1)$ by induction; explain the calculus analogy.

SOLUTION. Induction on n. For $n = 1$, the left side is $f(1) - f(2)$, matching $f(1) - f(2)$. Assume the identity for n. Then

$$\sum_{k=1}^{n+1} (f(k) - f(k+1)) = \underbrace{(f(1) - f(n+1))}_{\text{by hypothesis}} + (f(n+1) - f(n+2)) = f(1) - f(n+2),$$

the $f(n+1)$ terms cancelling. This is the claim for $n+1$, completing the induction.

The calculus analogy. A telescoping sum collapses a sum of consecutive *differences* to its endpoints, exactly as the Fundamental Theorem of Calculus collapses an integral of a *derivative* $\int_a^b F'(x)\,dx = F(b) - F(a)$ to boundary values. The difference $f(k) - f(k+1)$ is the discrete analogue of $-F'$, and the sum is the discrete integral. ■

Problem 11. Prove $\sum_{k=1}^{n} \arctan \frac{1}{k^2+k+1} = \arctan(n+1) - \arctan 1$.

SOLUTION. Recognize each term as a difference of arctangents. The subtraction formula is $\arctan a - \arctan b = \arctan \frac{a-b}{1+ab}$ (valid here, as all quantities are positive and the products keep us in range). Set $a = k+1$ and $b = k$:

$$\frac{a-b}{1+ab} = \frac{(k+1)-k}{1+k(k+1)} = \frac{1}{k^2+k+1}.$$

So $\arctan \frac{1}{k^2+k+1} = \arctan(k+1) - \arctan(k)$, and the sum telescopes:

$$\sum_{k=1}^{n} (\arctan(k+1) - \arctan(k)) = \arctan(n+1) - \arctan(1).$$

■

Problem 12. Using $\sum_{k=1}^{n} k^2 = \frac{n(n+1)(2n+1)}{6}$, show $\frac{1}{n^3}\sum_{k=1}^{n} k^2 \to \frac{1}{3}$, and explain why this equals $\int_0^1 x^2\,dx$.

SOLUTION. Divide by n^3 and read it as a Riemann sum. From the closed form,

$$\frac{1}{n^3}\sum_{k=1}^{n} k^2 = \frac{n(n+1)(2n+1)}{6n^3} = \frac{(n+1)(2n+1)}{6n^2} = \frac{2n^2+3n+1}{6n^2} \xrightarrow{n\to\infty} \frac{2}{6} = \frac{1}{3}.$$

The same quantity is a Riemann sum: $\frac{1}{n^3}\sum k^2 = \sum_{k=1}^{n}\left(\frac{k}{n}\right)^2 \cdot \frac{1}{n}$, which samples $f(x) = x^2$ at $x = \frac{k}{n}$ with rectangle width $\frac{1}{n}$. As $n \to \infty$ the width shrinks to 0 and the sum becomes

$$\int_0^1 x^2\,dx = \left[\frac{x^3}{3}\right]_0^1 = \frac{1}{3}.$$

This is precisely the telescoping-to-integral bridge: the discrete power sum, rescaled, converges to the integral of the corresponding power. ■

The Triangle That Runs Everything

Full worked solutions. Each one crosses the bridge between a coefficient and a count of subsets.

Launch

Problem 1. Find the coefficient of x^3 in $(1+x)^7$.

SOLUTION. By the binomial theorem the coefficient of x^3 is $\binom{7}{3} = 35$.

35

Problem 2. Find the coefficient of x^3 in $(2+x)^6$.

SOLUTION. Read the right term off the binomial theorem. By the binomial theorem,

$$(2+x)^6 = \sum_{k=0}^{6} \binom{6}{k} 2^{6-k} x^k,$$

so the term carrying x^3 is $\binom{6}{3}2^{6-3}x^3$. Now $\binom{6}{3} = 20$ and $2^3 = 8$, so the coefficient is $20 \cdot 8 = 160$. **The coefficient of x^3 is 160.**

160

Core

Problem 3. Prove $\sum_{k=0}^{n} \binom{n}{k} = 2^n$ two ways.

SOLUTION. Algebra. Set $a = b = 1$ in $(a+b)^n = \sum_k \binom{n}{k} a^{n-k} b^k$: the left side is 2^n, the right side is $\sum_k \binom{n}{k}$.

Counting. The right side counts all subsets of an n-element set, grouped by size. The total number of subsets is 2^n, since each element is independently in or out. Both counts give the same total, so $\sum_k \binom{n}{k} = 2^n$. ∎

Problem 4. Prove $\sum_{k=0}^{n} k\binom{n}{k} = n\,2^{n-1}$ by differentiating $(1+x)^n$.

SOLUTION. Differentiate the binomial expansion. Start from $(1+x)^n = \sum_{k=0}^{n} \binom{n}{k} x^k$. Differentiate both sides with respect to x:

$$n(1+x)^{n-1} = \sum_{k=0}^{n} k\binom{n}{k} x^{k-1}.$$

Now set $x = 1$: the left side becomes $n\,2^{n-1}$, and the right side becomes $\sum_k k\binom{n}{k}$. Hence $\sum_k k\binom{n}{k} = n\,2^{n-1}$. (This matches the committee-with-a-captain count of Chapter 1.) ∎

Problem 5. Evaluate $\binom{2}{2} + \binom{3}{2} + \cdots + \binom{10}{2}$ via the hockey stick.

SOLUTION. The hockey-stick identity gives $\sum_{i=2}^{10} \binom{i}{2} = \binom{11}{3}$. Computing, $\binom{11}{3} = \frac{11\cdot 10\cdot 9}{6} = 165$.

165

Stretch

Problem 6. Prove Vandermonde's identity $\sum_k \binom{m}{k}\binom{n}{r-k} = \binom{m+n}{r}$ by the two-rooms argument.

SOLUTION. Count an r-person committee from two rooms. Place m people in one room ("seniors") and n in another ("juniors"), $m+n$ in all. The number of r-person committees is $\binom{m+n}{r}$ directly. Alternatively, classify a committee by the number k of seniors it contains: choose k seniors in $\binom{m}{k}$ ways and the remaining $r-k$ members from the juniors in $\binom{n}{r-k}$ ways. Summing over k,

$$\binom{m+n}{r} = \sum_k \binom{m}{k}\binom{n}{r-k}.$$

∎

Problem 7. Find the last two digits of 3^{100} using $3^{100} = (10-1)^{50}$.

SOLUTION. Expand and keep only low powers of 10. Write $3^{100} = 9^{50} = (10-1)^{50} = \sum_k \binom{50}{k} 10^k (-1)^{50-k}$. Modulo 100, every term with $k \geq 2$ vanishes (divisible by 100). Keep $k = 0$ and $k = 1$:

$$(-1)^{50} + \binom{50}{1}(10)(-1)^{49} = 1 + 500\cdot(-1) = 1 - 500 \equiv 1 \pmod{100}.$$

The last two digits are 01.

01

Problem 8. Find the constant term in the expansion of $\left(x + \frac{2}{x}\right)^6$.

SOLUTION. Write the general term and kill the power of x. By the binomial theorem the general term is

$$\binom{6}{k} x^{6-k} \left(\frac{2}{x}\right)^k = \binom{6}{k} 2^k x^{6-k} x^{-k} = \binom{6}{k} 2^k x^{6-2k}.$$

The constant term has $x^{6-2k} = x^0$, i.e. $6 - 2k = 0$, so $k = 3$. Substituting,

$$\binom{6}{3} 2^3 = 20 \cdot 8 = 160.$$

The constant term is 160.

160

The Flash

Problem 9. Sum of all entries in rows 0 through n of Pascal's triangle?

SOLUTION. Row i sums to 2^i, so the total over rows 0 through n is $2^0 + 2^1 + \cdots + 2^n = 2^{n+1} - 1$.

$2^{n+1} - 1$

Problem 10. Coefficient of x^4 in $(1 + 2x)^5$?

SOLUTION. Track the power of 2. By the binomial theorem $(1+2x)^5 = \sum_k \binom{5}{k}(2x)^k = \sum_k \binom{5}{k}2^k x^k$. The x^4 term has $k = 4$: $\binom{5}{4}2^4 = 5 \cdot 16 = 80$.

80

Proof Lab

Problem 11. Prove the binomial theorem by a counting argument.

SOLUTION. Expand by choosing a term from each factor. Write $(a + b)^n = \underbrace{(a + b)(a + b) \cdots (a + b)}_{n}$. Expanding fully, each term of the product is formed by choosing either a or b from each of the n factors and multiplying the choices. A term equals $a^{n-k}b^k$ precisely when we chose b from exactly k of the factors and a from the other $n-k$. The number of ways to pick which k factors supply the b is $\binom{n}{k}$, so the term $a^{n-k}b^k$ appears $\binom{n}{k}$ times. Collecting,

$$(a+b)^n = \sum_{k=0}^{n} \binom{n}{k} a^{n-k} b^k.$$

■

Problem 12. Prove $\sum_{k=0}^{n} \binom{n}{k}^2 = \binom{2n}{n}$ and state the Vandermonde identity it specializes.

SOLUTION. Two-rooms committee with equal rooms. Choose an n-person committee from n seniors and n juniors — $\binom{2n}{n}$ ways directly. Classifying by the number k of seniors chosen gives $\sum_k \binom{n}{k}\binom{n}{n-k}$. Since $\binom{n}{n-k} = \binom{n}{k}$ (choosing which juniors are out is choosing which are in), this equals $\sum_k \binom{n}{k}^2$. Hence

$$\sum_{k=0}^{n} \binom{n}{k}^2 = \binom{2n}{n}.$$

This is the special case $m = n$, $r = n$ of **Vandermonde's identity** $\sum_k \binom{m}{k}\binom{n}{r-k} = \binom{m+n}{r}$. ■

A Clothesline for Numbers

Full worked solutions. Each one hangs a sequence on a power series and does algebra to extract a coefficient.

Launch

Problem 1. Closed forms for the generating functions of $1,1,1,\ldots$ and $1,2,3,4,\ldots$.

SOLUTION. The constant sequence gives the geometric series $\sum_{n\geq 0} x^n = \frac{1}{1-x}$. Differentiating $\frac{1}{1-x}$ gives $\sum_{n\geq 1} nx^{n-1} = \frac{1}{(1-x)^2}$, which is the generating function for $1,2,3,4,\ldots$

$\frac{1}{1-x}$ **and** $\frac{1}{(1-x)^2}$

Problem 2. Coefficient of x^4 in $(1+x+x^2+\cdots)^2$; interpret it.

SOLUTION. Squaring the geometric series,

$$(1+x+x^2+\cdots)^2 = \frac{1}{(1-x)^2} = \sum_n (n+1)x^n,$$

so the coefficient of x^4 is 5. Interpretation: it counts the ways to write $4 = i + j$ with $i,j \geq 0$ — namely $(0,4)$, $(1,3)$, $(2,2)$, $(3,1)$, $(4,0)$, which is 5.

Core

Problem 3. Number of ways to make 25 cents with pennies, nickels, dimes, quarters, via generating functions.

SOLUTION. One series per coin, then multiply. Each coin type contributes a factor recording the amounts it can supply: pennies $\frac{1}{1-x}$, nickels $\frac{1}{1-x^5}$, dimes $\frac{1}{1-x^{10}}$, quarters $\frac{1}{1-x^{25}}$. The number of ways to make 25 cents is the coefficient of x^{25} in

$$\frac{1}{(1-x)(1-x^5)(1-x^{10})(1-x^{25})}.$$

Expanding (by hand-tracking how many nickels/dimes/quarters are used, then filling with pennies) gives 13 ways. **There are 13 ways to make 25 cents.**

13

Problem 4. Generating function and closed form for $a_n = a_{n-1} + 2$, $a_0 = 1$.

SOLUTION. Translate the recurrence. The terms are $1, 3, 5, 7, \ldots$ — the odd numbers — so we expect $a_n = 2n + 1$. To see it via generating functions, let $A(x) = \sum_n a_n x^n$. The recurrence $a_n - a_{n-1} = 2$ for $n \geq 1$ gives $A(x) - xA(x) = a_0 + \sum_{n\geq 1} 2x^n = 1 + \frac{2x}{1-x}$, so

$$A(x) = \frac{1}{1-x} + \frac{2x}{(1-x)^2}.$$

Reading coefficients, $a_n = 1 + 2n$. **So $a_n = 2n + 1$.**

$a_n = 2n + 1$

Problem 5. Show compositions of n into 1s and 2s have generating function $\frac{1}{1-x-x^2}$; connect to Fibonacci.

SOLUTION. One factor per part, summed as a geometric series. A composition into parts of size 1 or 2 is an ordered sequence of such parts. Each part contributes either x^1 or x^2 to the exponent (its size), so a single part corresponds to $x + x^2$. A composition is any sequence of parts, and summing over all lengths (a geometric series in the "one part" generating function) gives

$$\sum_{\ell\geq 0} (x + x^2)^\ell = \frac{1}{1-(x+x^2)} = \frac{1}{1-x-x^2}.$$

The coefficient of x^n counts compositions of n; this is the Fibonacci generating function, so the count is F_{n+1}, matching the staircase-climbing problem of Chapter 5. ■

Stretch

Problem 6. Generating function for partitions of n into parts ≤ 3; compute for $n = 6$.

SOLUTION. One factor per allowed part size. A partition into parts of size at most 3 uses some number of 1s, 2s, and 3s. The choices are independent, and the count of each part size is unrestricted, so each size j contributes $\frac{1}{1-x^j}$:

$$\frac{1}{(1-x)(1-x^2)(1-x^3)}.$$

The coefficient of x^6 counts the partitions of 6 into parts ≤ 3:

$$3+3, \quad 3+2+1, \quad 3+1+1+1, \quad 2+2+2,$$

$$2+2+1+1, \quad 2+1+1+1+1, \quad 1+1+1+1+1+1,$$

seven of them. **The count is 7.**

7

Problem 7. Solve $a_n = 3a_{n-1} - 2a_{n-2}, a_0 = 0, a_1 = 1$, via partial fractions.

SOLUTION. Form the rational generating function. Let $A(x) = \sum_n a_n x^n$. The recurrence gives, after the standard shift-and-subtract,

$$A(x)(1-3x+2x^2) = a_0 + (a_1 - 3a_0)x = x,$$

so $A(x) = \frac{x}{1-3x+2x^2} = \frac{x}{(1-x)(1-2x)}$. Partial fractions:

$$\frac{x}{(1-x)(1-2x)} = \frac{-1}{1-x} + \frac{1}{1-2x} = \sum_n (2^n - 1)x^n.$$

Hence $a_n = 2^n - 1$.

$a_n = 2^n - 1$

The Flash

Problem 8. Coefficient of x^7 in $\frac{1}{(1-x)^3}$?

SOLUTION. The expansion is $\frac{1}{(1-x)^3} = \sum_n \binom{n+2}{2} x^n$, so the coefficient of x^7 is $\binom{9}{2} = 36$.

36

Problem 9. Generating function for $0, 1, 2, 3, \ldots$?

SOLUTION. This is $x \cdot \frac{1}{(1-x)^2} = \frac{x}{(1-x)^2}$ (shift the $1, 2, 3, \ldots$ series up by one place).

$\frac{x}{(1-x)^2}$

Proof Lab

Problem 10. Derive the Catalan generating function from $C = 1 + xC^2$, choosing the right branch.

SOLUTION. Solve the quadratic, then pick the branch by $C(0) = 1$. Treat $C = C(x)$ as the unknown in $xC^2 - C + 1 = 0$. The quadratic formula gives

$$C(x) = \frac{1 \pm \sqrt{1-4x}}{2x}.$$

We need $C(0) = C_0 = 1$. As $x \to 0$, the "+" branch behaves like $\frac{1+1}{2x} = \frac{1}{x} \to \infty$, which is impossible for a power series; the "−" branch behaves like $\frac{1-(1-2x+\cdots)}{2x} = \frac{2x+\cdots}{2x} \to 1$. So we take the minus sign:

$$C(x) = \frac{1 - \sqrt{1-4x}}{2x}.$$

Expanding $\sqrt{1-4x}$ by the binomial series and simplifying yields $C_n = \frac{1}{n+1}\binom{2n}{n}$. ■

Problem 11. Prove the Fibonacci generating function is $\frac{x}{1-x-x^2}$, justifying shift-equals-multiply.

SOLUTION. Multiplying a generating function by x shifts its coefficients up by one. Let $F(x) = \sum_{n\geq 0} F_n x^n$ with $F_0 = 0, F_1 = 1$. Because the coefficient of x^n in $xF(x)$ is F_{n-1} and in $x^2F(x)$ is F_{n-2}, the combination $F(x) - xF(x) - x^2F(x)$ has, for each $n \geq 2$, coefficient $F_n - F_{n-1} - F_{n-2} = 0$ by the recurrence. Only the low-order terms survive: the x^0 coefficient is $F_0 = 0$, and the x^1 coefficient is $F_1 - F_0 = 1$. Hence

$$F(x)(1 - x - x^2) = x \quad \Longrightarrow \quad F(x) = \frac{x}{1 - x - x^2}.$$

■

Only the Remainder Matters

Full worked solutions. Each one replaces numbers by remainders and keeps the arithmetic small.

Launch

Problem 1. Remainder of 1234 divided by 9.

SOLUTION. A number is congruent to its digit sum mod 9 (since $10 \equiv 1$). The digit sum of 1234 is $1 + 2 + 3 + 4 = 10 \equiv 1 \pmod{9}$. **The remainder is** 1.

1

Problem 2. Show $\overline{ababab}$ is divisible by 10101, and factor 10101.

SOLUTION. Repeating a two-digit block multiplies by 10101. Writing the two-digit block $\overline{ab}$ three times places copies at the ten-thousands and hundreds positions:

$$\overline{ababab} = \overline{ab} \cdot 10000 + \overline{ab} \cdot 100 + \overline{ab} = \overline{ab}\,(10000 + 100 + 1) = \overline{ab} \cdot 10101.$$

So $10101 \mid \overline{ababab}$ always. Factoring, $10101 = 3 \times 7 \times 13 \times 37$. ■

Problem 3. Find the inverse of 5 modulo 13.

SOLUTION. Hunt for the multiple of 5 that is one more than a multiple of 13. Since $\gcd(5, 13) = 1$, an inverse exists. Run through multiples of 5 mod 13: $5, 10, 15 \equiv 2, 20 \equiv 7, 25 \equiv 12, 30 \equiv 4, 35 \equiv 9, 40 \equiv 1$. So $5 \cdot 8 = 40 = 39 + 1 \equiv 1 \pmod{13}$. **The inverse is** $5^{-1} \equiv 8 \pmod{13}$.

8

Core

Problem 4. Last digit of 7^{2024} via the cycle of 7^k mod 10.

SOLUTION. Find the cycle. The powers of 7 mod 10 run $7^1 \equiv 7, 7^2 \equiv 9, 7^3 \equiv 3, 7^4 \equiv 1$, then repeat with period 4. Reduce the exponent mod 4: $2024 \equiv 0 \pmod 4$, which corresponds to the end of a cycle, $7^4 \equiv 1$. **The last digit is** 1.

1

Problem 5. Prove $6 \mid n^3 - n$ for all integers n.

SOLUTION. Factor into consecutive integers. We have $n^3 - n = n(n^2-1) = (n-1)n(n+1)$, the product of three consecutive integers. Among any three consecutive integers, at least one is even (so 2 divides the product) and exactly one is a multiple of 3 (so 3 divides it). Since 2 and 3 are coprime, their product 6 divides $n^3 - n$. ■

Problem 6. Derive the divisibility-by-11 rule.

SOLUTION. Reduce powers of 10 mod 11. Since $10 \equiv -1 \pmod{11}$, we get $10^k \equiv (-1)^k$: the powers alternate $1, -1, 1, -1, \ldots$ Writing a number by its digits, $\overline{d_k \cdots d_1 d_0} = \sum_i d_i 10^i \equiv \sum_i d_i (-1)^i \pmod{11}$. So a number is congruent mod 11 to its *alternating digit sum* $d_0 - d_1 + d_2 - \cdots$, and is divisible by 11 exactly when that alternating sum is. ■

Problem 7. Solve $3x \equiv 4 \pmod 7$ using an inverse.

SOLUTION. Multiply by 3^{-1} to undo the multiplication. Since $\gcd(3,7) = 1$, the inverse exists; checking multiples of 3, we find $3 \cdot 5 = 15 \equiv 1 \pmod 7$, so $3^{-1} \equiv 5$. Multiply both sides of $3x \equiv 4$ by 5:

$$x \equiv 5 \cdot 4 = 20 \equiv 6 \pmod 7.$$

Check: $3 \cdot 6 = 18 = 14 + 4 \equiv 4 \pmod 7$. **So** $x \equiv 6 \pmod 7$.

$$x \equiv 6 \pmod 7$$

Stretch

Problem 8. Remainder of 3^{100} divided by 100.

SOLUTION. Find the cycle of 3^k mod 100. The powers of 3 modulo 100 repeat with period 20 (one checks $3^{20} \equiv 1 \pmod{100}$). Reduce the exponent: $100 = 5 \cdot 20$, so $3^{100} = (3^{20})^5 \equiv 1^5 = 1 \pmod{100}$. **The last two digits are** 01.

01

Problem 9. Solve $x \equiv 2 \pmod 3$ and $x \equiv 3 \pmod 5$.

SOLUTION. Combine via the Chinese Remainder Theorem. Since $\gcd(3,5) = 1$, there is a unique solution mod 15. List the residues $\equiv 3 \pmod 5$: $3, 8, 13$; of these, the one $\equiv 2 \pmod 3$ is 8 (as $8 = 3 \cdot 2 + 2$). So $x \equiv 8 \pmod{15}$. **All solutions are** $x \equiv 8 \pmod{15}$.

$x \equiv 8 \pmod{15}$

Problem 10. List all nonzero quadratic residues mod 11, and check there are 5.

SOLUTION. Square every residue and reduce. Because x and $-x$ square to the same value, we need only square 1 through 5 (then the rest mirror):

$$1^2 \equiv 1, \quad 2^2 \equiv 4, \quad 3^2 \equiv 9, \quad 4^2 = 16 \equiv 5, \quad 5^2 = 25 \equiv 3 \pmod{11}.$$

The residues $6, 7, 8, 9, 10$ (which are $-5, -4, -3, -2, -1$) give the same outputs $3, 5, 9, 4, 1$. So the distinct quadratic residues are

$$\{1, 3, 4, 5, 9\},$$

exactly $(11-1)/2 = 5$ of them; the non-residues are $\{2, 6, 7, 8, 10\}$. ■

$\{1, 3, 4, 5, 9\}$

The Flash

Problem 11. Compute $5^{99} \bmod 7$.

SOLUTION. Since $5 \equiv -2 \pmod 7$, note $5^3 = 125 \equiv -1 \pmod 7$ (indeed $125 = 18 \cdot 7 - 1$). So the powers of 5 have order 6, but more usefully $5^3 \equiv -1$. Writing $99 = 3 \cdot 33$, $5^{99} = (5^3)^{33} \equiv (-1)^{33} = -1 \equiv 6 \pmod 7$.

6

Problem 12. Is 918082 divisible by 11?

SOLUTION. Compute the alternating digit sum (from the units digit): $2 - 8 + 0 - 8 + 1 - 9 = -22$. Since $11 \mid -22$, yes, 918082 is divisible by 11.

Yes

Proof Lab

Problem 13. Prove $30 \mid n^5 - n$ for every integer n.

SOLUTION. Split $30 = 2 \cdot 3 \cdot 5$ and check each prime. Factor $n^5 - n = (n-1)n(n+1)(n^2+1)$.

Divisible by 2*:* among the consecutive integers $n-1, n, n+1$ at least one is even.

Divisible by 3*:* among three consecutive integers one is a multiple of 3.

Divisible by 5*:* work mod 5. Checking the five residues, $0^5 \equiv 0$, $1^5 \equiv 1$, $2^5 = 32 \equiv 2$, $3^5 = 243 \equiv 3$, $4^5 = 1024 \equiv 4$; in every case $n^5 \equiv n \pmod 5$, so $5 \mid n^5 - n$.

Since $2, 3, 5$ are pairwise coprime and each divides $n^5 - n$, their product 30 divides it. ■

Problem 14. Prove $\overline{abcabc} = \overline{abc} \times 1001$, conclude divisibility by $7, 11, 13$, and give the four-digit analogue.

SOLUTION. Repeating a three-digit block multiplies by 1001. Placing $\overline{abc}$ in the thousands and ones positions,

$$\overline{abcabc} = \overline{abc} \cdot 1000 + \overline{abc} = \overline{abc}(1000+1) = \overline{abc} \cdot 1001.$$

Since $1001 = 7 \times 11 \times 13$, every such number is divisible by 7, 11, and 13.

Four-digit analogue. A four-digit block $\overline{abcd}$ repeated twice gives

$$\overline{abcdabcd} = \overline{abcd} \cdot 10000 + \overline{abcd} = \overline{abcd}\,(10001),$$

and $10001 = 73 \times 137$, so such numbers are always divisible by 73 and 137. ■

Problem 15. Prove that no perfect square ends in the digit 2, 3, 7, or 8.

SOLUTION. The last digit is a mod-10 fact. The last digit of n^2 depends only on the last digit of n, so it is enough to square 0 through 9:

$$0,\ 1,\ 4,\ 9,\ 16,\ 25,\ 36,\ 49,\ 64,\ 81 \ \longrightarrow\ \text{last digits } 0,1,4,9,6,5,6,9,4,1.$$

Every perfect square ends in one of $\{0,1,4,5,6,9\}$; the digits $2,3,7,8$ never appear. ■

Problem 16. Show that $x^2 - 5y^2 = 2$ has no integer solutions.

SOLUTION. Reduce mod 5 to isolate x^2. Modulo 5 the term $5y^2$ vanishes, so the equation becomes $x^2 \equiv 2 \pmod 5$. But the squares mod 5 are $0^2, 1^2, 2^2, 3^2, 4^2 \equiv 0, 1, 4, 4, 1$, i.e. only $\{0,1,4\}$ — and 2 is not among them. So $x^2 \equiv 2 \pmod 5$ is impossible, and the equation has no integer solutions. ■

Problem 17. Prove that $x^3 + y^3 = 3$ has no integer solutions.

SOLUTION. Cubes are tightly constrained mod 9. Checking $0^3, \dots, 8^3$ modulo 9 gives only 0, 1, and $8 \equiv -1$, so every cube is 0 or $\pm 1 \pmod 9$. A sum of two cubes can therefore only be

$$x^3 + y^3 \equiv \{0, \pm 1, \pm 2\} = \{0, 1, 2, 7, 8\} \pmod 9,$$

which omits 3. Since $3 \equiv 3 \pmod 9$, no integers x, y can satisfy $x^3 + y^3 = 3$. ∎

When Giants Collapse

Full worked solutions. Each one lets a theorem hand over the cycle length, then reduces the exponent.

Launch

Problem 1. Compute $\varphi(12)$, $\varphi(13)$, $\varphi(36)$ by the product formula.

SOLUTION. Using $\varphi(m) = m\prod_{p|m}(1-\frac{1}{p})$:

$$\varphi(12) = 12\big(1-\tfrac{1}{2}\big)\big(1-\tfrac{1}{3}\big) = 12\cdot\tfrac{1}{2}\cdot\tfrac{2}{3} = 4,$$

$\varphi(13) = 13 - 1 = 12$ (prime), and $\varphi(36) = 36\cdot\frac{1}{2}\cdot\frac{2}{3} = 12$.

$$\varphi(12) = 4,\ \varphi(13) = 12,\ \varphi(36) = 12$$

Problem 2. Find 3^6 mod 7 using Fermat.

SOLUTION. Fermat's Little Theorem says $a^{p-1} \equiv 1 \pmod{p}$ for prime $p \nmid a$. With $p = 7$ and $a = 3$, $3^6 \equiv 1 \pmod{7}$.

Problem 3. Find the order of 2 mod 7 by listing its powers until they return to 1.

SOLUTION. The order is the first exponent giving 1. List the powers of 2 mod 7: $2^1 \equiv 2$, $2^2 \equiv 4$, $2^3 \equiv 8 \equiv 1$. The first return to 1 is at exponent 3, so the order of 2 is 3. (As a check, $3 \mid \varphi(7) = 6$, as the order-divides-φ theorem requires. Unlike 3, the element 2 is *not* a primitive root mod 7 — its powers cycle through only $\{1, 2, 4\}$.)

$$\textbf{order} = 3$$

Core

Problem 4. Find 3^{100} mod 7 by reducing the exponent.

SOLUTION. By Fermat $3^6 \equiv 1 \pmod 7$, so powers of 3 repeat every 6 steps. Reduce the exponent: $100 = 6 \cdot 16 + 4$, so $3^{100} \equiv 3^4 \pmod 7$. Now $3^4 = 81 = 7 \cdot 11 + 4 \equiv 4$. **So** $3^{100} \equiv 4 \pmod 7$.

4

Problem 5. Find the order of 3 mod 7; verify it divides $\varphi(7) = 6$.

SOLUTION. Compute powers until reaching 1. The powers of 3 mod 7 are $3^1 \equiv 3$, $3^2 \equiv 2$, $3^3 \equiv 6$, $3^4 \equiv 4$, $3^5 \equiv 5$, $3^6 \equiv 1$. The first return to 1 is at exponent 6, so the order of 3 is 6. Indeed $6 \mid \varphi(7) = 6$, as the order-divides-φ theorem requires. (Here 3 is a *primitive root* mod 7, hitting all nonzero residues.)

order $= 6$

Problem 6. Last two digits of 3^{2024} using Euler mod 100.

SOLUTION. Euler gives the cycle length $\varphi(100) = 40$. Since $\gcd(3, 100) = 1$, Euler's theorem gives $3^{40} \equiv 1 \pmod{100}$. Reduce the exponent mod 40: $2024 = 40 \cdot 50 + 24$, so $3^{2024} \equiv 3^{24} \pmod{100}$. Computing 3^{24} mod 100 by repeated squaring gives 81. **The last two digits are** 81.

81

Stretch

Problem 7. Prove $\binom{p}{k} \equiv 0 \pmod p$ for $0 < k < p$ prime, and deduce Fermat's Little Theorem.

SOLUTION. A prime in the numerator has nothing to cancel it. Write $\binom{p}{k} = \frac{p!}{k!(p-k)!} = \frac{p\,(p-1)!}{k!(p-k)!}$. The numerator contains the prime factor p. For $0 < k < p$, both $k!$ and $(p-k)!$ are products of integers strictly less than p, so neither is divisible by p; since $\binom{p}{k}$ is an integer, the factor p survives, giving $p \mid \binom{p}{k}$.

Slick Fermat. Expand $a^p = (\underbrace{1 + 1 + \cdots + 1}_{a})^p$ by the multinomial theorem. Modulo p, every mixed term carries a multinomial coefficient divisible by p (same argument), leaving only the "pure" terms 1^p, one per summand: $a^p \equiv \underbrace{1 + \cdots + 1}_{a} = a \pmod p$. Hence $a^p \equiv a$, and dividing by a when $p \nmid a$ gives $a^{p-1} \equiv 1$. ■

Problem 8. Prove $a^{561} \equiv a \pmod{561}$ for all a, with $561 = 3 \times 11 \times 17$.

SOLUTION. Check each prime factor, then recombine by CRT. It suffices to show $a^{561} \equiv a$ modulo each of 3, 11, 17; the Chinese Remainder Theorem then gives it mod their product 561.

Take the prime $p \in \{3, 11, 17\}$. If $p \mid a$, both sides are $\equiv 0$. Otherwise Fermat gives $a^{p-1} \equiv 1 \pmod{p}$. The key arithmetic fact is that $p-1$ divides $560 = 561-1$ for each: $2 \mid 560$, $10 \mid 560$, and $16 \mid 560$ (indeed $560 = 16 \cdot 35$). So $a^{560} = (a^{p-1})^{560/(p-1)} \equiv 1 \pmod{p}$, hence $a^{561} = a \cdot a^{560} \equiv a \pmod{p}$. Holding for all three primes, $a^{561} \equiv a \pmod{561}$. This is why 561 is a *Carmichael number*: it passes Fermat's test $a^{560} \equiv 1$ for *every* base a coprime to it, mimicking a prime without being one — exactly the kind of composite that fools the Fermat primality test. ■

The Flash

Problem 9. Compute $7^{2026} \bmod 100$.

SOLUTION. $\gcd(7, 100) = 1$ and $\varphi(100) = 40$, so $7^{40} \equiv 1 \pmod{100}$. Reduce: $2026 \equiv 26 \pmod{40}$, and computing $7^{26} \bmod 100$ gives 49.

49

Problem 10. Compute 6! mod 7 and 10! mod 11 using Wilson.

SOLUTION. Wilson's theorem: $(p-1)! \equiv -1 \pmod{p}$ for prime p. So $6! \equiv -1 \equiv 6 \pmod{7}$ and $10! \equiv -1 \equiv 10 \pmod{11}$.

$6! \equiv 6,\ 10! \equiv 10$

Proof Lab

Problem 11. Prove the totient product formula $\varphi(m) = m \prod_{p \mid m} (1 - \frac{1}{p})$.

SOLUTION. Remove each prime's multiples; independence is CRT. Let $m = p_1^{a_1} \cdots p_r^{a_r}$. An integer in $\{1, \dots, m\}$ is coprime to m exactly when it is divisible by *none* of the primes $p_1, \dots, p_r$. Among $1, \dots, m$, a fraction $\frac{1}{p_i}$ are divisible by p_i, so a fraction $1 - \frac{1}{p_i}$ avoid p_i. The divisibility conditions for distinct primes are *independent* — by the Chinese Remainder Theorem, residues mod p_i and mod p_j can be chosen freely and independently — so the surviving fractions multiply:

$$\varphi(m) = m \prod_{i=1}^{r} \left(1 - \frac{1}{p_i}\right).$$

■

Problem 12. Prove Wilson's theorem $(p-1)! \equiv -1 \pmod{p}$.

SOLUTION. Pair each residue with its inverse. Work modulo a prime p. Each of $1, 2, \ldots, p-1$ has a unique multiplicative inverse among them. Pair each element with its inverse; a pair $\{a, a^{-1}\}$ with $a \neq a^{-1}$ contributes $a \cdot a^{-1} = 1$ to the product $(p-1)!$.

The only *self-inverse* elements satisfy $a^2 \equiv 1$, i.e. $(a-1)(a+1) \equiv 0 \pmod{p}$; since p is prime this forces $a \equiv 1$ or $a \equiv -1$. So in the full product $(p-1)!$, every element except 1 and $-1 \equiv p-1$ cancels in inverse-pairs, leaving

$$(p-1)! \equiv 1 \cdot (-1) = -1 \pmod{p}.$$

■

The Honeybee's Craft

Full worked solutions. Each one finds the bound and — just as important — the exact moment equality holds.

Launch

Problem 1. Among rectangles with perimeter 36, which has the largest area?

SOLUTION. Fixed sum, maximize product. The sides satisfy $a + b = 18$, and the area is ab. By AM-GM, $\sqrt{ab} \le \frac{a+b}{2} = 9$, so $ab \le 81$, with equality when $a = b = 9$. **The** 9×9 **square, area** 81.

9×9, **area** 81

Problem 2. Prove $x + \frac{1}{x} \ge 2$ for $x > 0$; equality?

SOLUTION. Apply AM-GM to the two positive terms x and $\frac{1}{x}$, whose product is 1:

$$\frac{x + \frac{1}{x}}{2} \ge \sqrt{x \cdot \tfrac{1}{x}} = 1 \quad \Longrightarrow \quad x + \tfrac{1}{x} \ge 2.$$

Equality holds when $x = \frac{1}{x}$, i.e. $x = 1$. ■

Core

Problem 3. Minimum of $x + \frac{9}{x}$ for $x > 0$.

SOLUTION. The product of the two terms is constant. Here $x \cdot \frac{9}{x} = 9$, a constant, so AM-GM gives a useful bound:

$$x + \frac{9}{x} \ge 2\sqrt{x \cdot \tfrac{9}{x}} = 2\sqrt{9} = 6,$$

with equality when $x = \frac{9}{x}$, i.e. $x^2 = 9$, $x = 3$. **Minimum** 6 **at** $x = 3$.

6 **at** $x = 3$

Problem 4. For positive a, b, c with $a + b + c = 1$, prove $\frac{1}{a} + \frac{1}{b} + \frac{1}{c} \geq 9$.

SOLUTION. Pair the sum with its reciprocal sum. By the Cauchy–Schwarz inequality (or AM-HM),

$$(a + b + c)\left(\frac{1}{a} + \frac{1}{b} + \frac{1}{c}\right) \geq (1 + 1 + 1)^2 = 9.$$

Since $a + b + c = 1$, this gives $\frac{1}{a} + \frac{1}{b} + \frac{1}{c} \geq 9$, with equality when $a = b = c = \frac{1}{3}$. ■

Problem 5. Use Cauchy–Schwarz to prove $(a + b + c)^2 \leq 3(a^2 + b^2 + c^2)$; equality?

SOLUTION. Apply Cauchy–Schwarz with the all-ones vector. Take $\vec{u} = (a, b, c)$ and $\vec{v} = (1, 1, 1)$. Cauchy–Schwarz says $(\vec{u} \cdot \vec{v})^2 \leq |\vec{u}|^2 |\vec{v}|^2$:

$$(a + b + c)^2 \leq (a^2 + b^2 + c^2)(1^2 + 1^2 + 1^2) = 3(a^2 + b^2 + c^2).$$

Equality holds when $\vec{u}$ and $\vec{v}$ are proportional, i.e. $a = b = c$. ■

Stretch

Problem 6. Maximize xyz subject to $x + y + z = 12$, $x, y, z > 0$.

SOLUTION. AM-GM on three variables. The arithmetic-geometric mean inequality gives

$$\frac{x + y + z}{3} \geq \sqrt[3]{xyz} \quad \Longrightarrow \quad \frac{12}{3} = 4 \geq \sqrt[3]{xyz} \quad \Longrightarrow \quad xyz \leq 64,$$

with equality when $x = y = z = 4$. **The maximum is** 64**, at the "cube"** $x = y = z = 4$**.**

64 **at** $x = y = z = 4$

Problem 7. Prove $a^2 + b^2 + c^2 \geq ab + bc + ca$ for all reals by forcing a sum of squares.

SOLUTION. Build non-negative squares. Consider

$$a^2 + b^2 + c^2 - ab - bc - ca = \frac{1}{2}\left((a - b)^2 + (b - c)^2 + (c - a)^2\right).$$

(Expand the right side: each square contributes, and the cross terms reassemble the left side.) A sum of squares is ≥ 0, so $a^2 + b^2 + c^2 \geq ab + bc + ca$, with equality exactly when $a = b = c$. ■

Problem 8. For positive a, b, c, prove $\frac{a^2}{b+c} + \frac{b^2}{c+a} + \frac{c^2}{a+b} \geq \frac{a+b+c}{2}$; equality?

SOLUTION. Titu's lemma (Cauchy–Schwarz in Engel form). For real a_i and positive b_i, Titu's lemma states

$$\frac{a_1^2}{b_1} + \cdots + \frac{a_n^2}{b_n} \geq \frac{(a_1 + \cdots + a_n)^2}{b_1 + \cdots + b_n}.$$

Apply it with numerators a, b, c and denominators $b+c,\ c+a,\ a+b$ (all positive):

$$\frac{a^2}{b+c} + \frac{b^2}{c+a} + \frac{c^2}{a+b} \geq \frac{(a+b+c)^2}{(b+c)+(c+a)+(a+b)} = \frac{(a+b+c)^2}{2(a+b+c)} = \frac{a+b+c}{2}.$$

Equality in Titu's lemma needs $\frac{a}{b+c} = \frac{b}{c+a} = \frac{c}{a+b}$, which holds exactly when $a = b = c$. ■

$\geq \frac{a+b+c}{2}$, **equality at** $a = b = c$

The Flash

Problem 9. Minimum of $\frac{a}{b} + \frac{b}{a}$ for positive a, b.

SOLUTION. The two terms have product $\frac{a}{b} \cdot \frac{b}{a} = 1$, so by AM-GM their sum is $\geq 2\sqrt{1} = 2$, with equality when $\frac{a}{b} = \frac{b}{a}$, i.e. $a = b$. **Minimum** 2.

2 **at** $a = b$

Problem 10. For $a = 4, b = 9$ compute QM, AM, GM, HM and confirm the ordering.

SOLUTION.

$$\text{QM} = \sqrt{\tfrac{16+81}{2}} = \sqrt{48.5} \approx 6.96, \qquad \text{AM} = \tfrac{13}{2} = 6.5,$$

$$\text{GM} = \sqrt{36} = 6, \qquad \text{HM} = \frac{2}{\frac{1}{4}+\frac{1}{9}} = \tfrac{72}{13} \approx 5.54.$$

These are strictly descending: $6.96 > 6.5 > 6 > 5.54$, confirming QM $\geq$ AM $\geq$ GM $\geq$ HM.

$6.96 > 6.5 > 6 > 5.54$

Proof Lab

Problem 11. Prove two-variable AM-GM by forcing a square, then the full ladder QM $\geq$ AM $\geq$ GM $\geq$ HM.

SOLUTION. AM $\geq$ GM. For $a, b \geq 0$,

$$\frac{a+b}{2} - \sqrt{ab} = \frac{1}{2}(a - 2\sqrt{ab} + b) = \frac{1}{2}(\sqrt{a} - \sqrt{b})^2 \geq 0,$$

with equality iff $\sqrt{a} = \sqrt{b}$, i.e. $a = b$.

QM $\geq$ AM. Squaring, this is $\frac{a^2+b^2}{2} \geq \left(\frac{a+b}{2}\right)^2$, which rearranges to $\frac{1}{4}(a-b)^2 \geq 0$ — true, equality iff $a = b$.

GM $\geq$ HM. The harmonic mean is $\frac{2ab}{a+b}$. The claim $\sqrt{ab} \geq \frac{2ab}{a+b}$ rearranges (multiply by $\frac{a+b}{\sqrt{ab}} > 0$) to $a + b \geq 2\sqrt{ab}$ — which is AM-GM again, equality iff $a = b$.

Chaining the three links gives QM $\geq$ AM $\geq$ GM $\geq$ HM, and every inequality is an equality precisely when $a = b$, so the ladder collapses to a point exactly then. ∎

Problem 12. Prove the n-variable AM-GM and identify the equality case.

SOLUTION. Smoothing argument. Fix the sum $S = x_1 + \cdots + x_n$ and consider maximizing the product $P = x_1 \cdots x_n$ over non-negative values with that sum. If two of the values differ, say $x_i < x_j$, replace both by their average $m = \frac{x_i + x_j}{2}$. This keeps the sum fixed, but the product of those two factors strictly increases, since

$$m^2 = \left(\frac{x_i + x_j}{2}\right)^2 > x_i x_j \quad (\text{equivalent to } (x_i - x_j)^2 > 0),$$

so P strictly increases. Therefore P is maximized only when no two values differ — all equal to $\frac{S}{n}$. At that point $P = \left(\frac{S}{n}\right)^n$, i.e. $\sqrt[n]{P} = \frac{S}{n}$, giving $\frac{x_1 + \cdots + x_n}{n} \geq \sqrt[n]{x_1 \cdots x_n}$ for all configurations, with equality iff $x_1 = \cdots = x_n$. ∎

Problem 13. State and prove the rearrangement inequality by the swap argument, and use it to prove $\frac{a}{b} + \frac{b}{c} + \frac{c}{a} \geq 3$ for positive a, b, c.

SOLUTION. The statement. Let $a_1 \leq \cdots \leq a_n$ and $b_1 \leq \cdots \leq b_n$ be sorted, and let σ be any permutation. Then

$$\sum_i a_i b_{n+1-i} \leq \sum_i a_i b_{\sigma(i)} \leq \sum_i a_i b_i,$$

i.e. the paired-product sum is largest when both lists are sorted the same way and smallest when sorted oppositely.

The swap argument. Suppose a pairing is not fully same-sorted: then for some i, j with $a_i < a_j$ the partners are out of order, $b_p > b_q$ with $a_i \mapsto b_p$, $a_j \mapsto b_q$. Swapping their partners changes the sum by

$$(a_i b_q + a_j b_p) - (a_i b_p + a_j b_q) = (a_j - a_i)(b_p - b_q) \geq 0,$$

a product of two non-negative factors. Each such swap moves the pairing toward same-sorted without ever decreasing the sum, so the same-sorted pairing is maximal; the mirror argument (every swap toward oppositely-sorted never increases the sum) gives the minimum.

Application. Pair the list (a, b, c) against the list of reciprocals $(\frac{1}{a}, \frac{1}{b}, \frac{1}{c})$. The key observation is that pairing each number with *its own* reciprocal,

$$a \cdot \tfrac{1}{a} + b \cdot \tfrac{1}{b} + c \cdot \tfrac{1}{c} = 1 + 1 + 1 = 3,$$

is the *oppositely-sorted* pairing: a larger number always carries a smaller reciprocal, so big is matched with small. By rearrangement, the oppositely-sorted pairing is the *minimum*, so every other permutation gives a sum that is at least 3. The cyclic permutation gives $a \cdot \frac{1}{b} + b \cdot \frac{1}{c} + c \cdot \frac{1}{a} = \frac{a}{b} + \frac{b}{c} + \frac{c}{a}$, hence

$$\frac{a}{b} + \frac{b}{c} + \frac{c}{a} \geq 3,$$

with equality when $a = b = c$ (so that all reciprocals coincide and every pairing gives the same sum).
■

The Back Door to the Imaginary

Full worked solutions. Each rides the two-way street between an equation and its roots: guess the shape, or read the cubic off the radical.

Launch

Problem 1. Solve $x^3 + 3x - 14 = 0$ by Cardano's formula, then denest.

SOLUTION. Here $p = 3$, $q = -14$, so $\Delta_0 = \frac{q^2}{4} + \frac{p^3}{27} = 49 + 1 = 50$ and $\sqrt{50} = 5\sqrt{2}$. Cardano gives

$$x = \sqrt[3]{7 + 5\sqrt{2}} + \sqrt[3]{7 - 5\sqrt{2}}.$$

Denest by guessing $\sqrt[3]{7 + 5\sqrt{2}} = a + b\sqrt{2}$. Cubing, $(a + b\sqrt{2})^3 = (a^3 + 6ab^2) + (3a^2b + 2b^3)\sqrt{2}$, so we need $a^3 + 6ab^2 = 7$ and $3a^2b + 2b^3 = 5$. Then $a = b = 1$ works, giving $\sqrt[3]{7 + 5\sqrt{2}} = 1 + \sqrt{2}$ and (same computation) $\sqrt[3]{7 - 5\sqrt{2}} = 1 - \sqrt{2}$. Hence $x = 2$; indeed $8 + 6 - 14 = 0$.

$$x = 2.$$

Problem 2. Solve $x^3 - 6x - 40 = 0$ by Cardano's formula and denest.

SOLUTION. Here $p = -6$, $q = -40$, so $\Delta_0 = 400 - 8 = 392$ and $\sqrt{392} = 14\sqrt{2}$. Then

$$x = \sqrt[3]{20 + 14\sqrt{2}} + \sqrt[3]{20 - 14\sqrt{2}}.$$

Guess $\sqrt[3]{20 + 14\sqrt{2}} = a + b\sqrt{2}$: we need $a^3 + 6ab^2 = 20$ and $3a^2b + 2b^3 = 14$, solved by $a = 2$, $b = 1$. So the two cube roots are $2 + \sqrt{2}$ and $2 - \sqrt{2}$, and $x = 4$ (check $64 - 24 - 40 = 0$).

$$x = 4.$$

Core

Problem 3. Solve $x^3 - 6x^2 + 6x - 2 = 0$ exactly.

SOLUTION. Depress with $x = t + 2$ (since $-a/3 = 2$):

$$(t+2)^3 - 6(t+2)^2 + 6(t+2) - 2 = t^3 - 6t - 6 = 0.$$

Now $p = q = -6$, $\Delta_0 = 9 - 8 = 1$, so $t = \sqrt[3]{3+1} + \sqrt[3]{3-1} = \sqrt[3]{4} + \sqrt[3]{2}$. The cubic $t^3 - 6t - 6$ has no rational root (none of $\pm1, \pm2, \pm3, \pm6$ works), so the cube roots do not denest. Restoring,

$$x = 2 + \sqrt[3]{2} + \sqrt[3]{4}.$$

$$x = 2 + \sqrt[3]{2} + \sqrt[3]{4}.$$

Problem 4. Simplify $\sqrt[3]{2+\sqrt{5}} + \sqrt[3]{2-\sqrt{5}}$.

SOLUTION. Let x be the value, with $A = 2 + \sqrt{5}$, $B = 2 - \sqrt{5}$. Then $s = A + B = 4$ and $AB = 4 - 5 = -1$, so $P = \sqrt[3]{AB} = -1$. Cubing $x = \sqrt[3]{A} + \sqrt[3]{B}$ gives $x^3 = s + 3Px = 4 - 3x$, i.e.

$$x^3 + 3x - 4 = 0.$$

Since $x^3 + 3x$ is strictly increasing, the rational root $x = 1$ $(1 + 3 - 4 = 0)$ is the only real root, so the expression equals 1.

1.

Problem 5. Simplify $\sqrt[3]{5+2\sqrt{13}} + \sqrt[3]{5-2\sqrt{13}}$.

SOLUTION. Here $s = 10$ and $AB = 25 - 4\cdot 13 = -27$, so $P = -3$. Thus $x^3 = 10 - 9x$, i.e. $x^3 + 9x - 10 = 0$. The strictly increasing cubic has the single real root $x = 1$ $(1 + 9 - 10 = 0)$.

1.

Stretch

Problem 6. Prove that $\sqrt[3]{\sqrt{5}+2} - \sqrt[3]{\sqrt{5}-2} = 1$.

SOLUTION. Set $a = \sqrt[3]{\sqrt{5}+2}$, $b = \sqrt[3]{\sqrt{5}-2}$, and $x = a-b$. Cubing a *difference*, $x^3 = a^3 - b^3 - 3ab(a-b) = a^3 - b^3 - 3ab\,x$. Now

$$a^3 - b^3 = (\sqrt{5}+2) - (\sqrt{5}-2) = 4, \qquad ab = \sqrt[3]{(\sqrt{5}+2)(\sqrt{5}-2)} = \sqrt[3]{5-4} = 1,$$

so $x^3 = 4 - 3x$, i.e. $x^3 + 3x - 4 = 0$. The only real root of this increasing cubic is $x = 1$. Hence the difference equals 1.

$= 1.$

Problem 7. Simplify $\sqrt[3]{26+15\sqrt{3}} + \sqrt[3]{26-15\sqrt{3}}$, first explaining the signs.

SOLUTION. **Signs first.** $15\sqrt{3} \approx 25.98 < 26$, so $26 - 15\sqrt{3} > 0$; both radicands are positive, hence both real cube roots are positive (no sign trap here). **Denest.** Guess $\sqrt[3]{26+15\sqrt{3}} = a + b\sqrt{3}$; cubing, $(a+b\sqrt{3})^3 = (a^3+9ab^2) + (3a^2b+3b^3)\sqrt{3}$, so $a^3 + 9ab^2 = 26$ and $3a^2b + 3b^3 = 15$. Then $a = 2, b = 1$ works: the cube roots are $2+\sqrt{3}$ and $2-\sqrt{3}$ (both positive), summing to 4.

4.

The Flash

Problem 8. Find all positive integers n for which $\sqrt[3]{n+\sqrt{n^2+8}} + \sqrt[3]{n-\sqrt{n^2+8}}$ is an integer.

SOLUTION. With $s = 2n$ and $AB = n^2 - (n^2+8) = -8$, we get $P = -2$, so the value x satisfies $x^3 + 6x - 2n = 0$, i.e. $x^3 + 6x = 2n$. If $x = k$ is an integer, then $k^3 + 6k = 2n$. Since $k^3 + 6k \equiv k \pmod{2}$, the left side is even only when k is even. Writing $k = 2m$,

$$2n = 8m^3 + 12m \quad \Longrightarrow \quad n = 4m^3 + 6m.$$

So the integer values occur exactly for $n = 4m^3 + 6m$ with $m = 1, 2, 3, \ldots$, namely $n = 10, 44, 126, 264, \ldots$ ($m = 1$ recovers the Opening Challenge, $n = 10$, value 2).

$n = 4m^3 + 6m$ **for** $m = 1, 2, 3, \ldots$ $(n = 10, 44, 126, \ldots)$.

Proof

Problem 9. Find the monic integer cubic with root $1 + \sqrt[3]{2}$.

SOLUTION. Let $x = 1 + \sqrt[3]{2}$, so $x - 1 = \sqrt[3]{2}$ and $(x-1)^3 = 2$. Expanding,

$$x^3 - 3x^2 + 3x - 1 = 2 \quad \Longrightarrow \quad x^3 - 3x^2 + 3x - 3 = 0.$$

This is monic with integer coefficients, and it is irreducible over $\mathbb{Q}$ (no rational root: $\pm 1, \pm 3$ all fail), so it is the minimal polynomial of $1 + \sqrt[3]{2}$.

$$x^3 - 3x^2 + 3x - 3 = 0.$$

Problem 10. Roots of $x^3 - 7x + 6 = 0$; then Cardano's assertion about $x = 1$.

SOLUTION. By inspection $x = 1, 2, -3$ are roots ($1 - 7 + 6 = 0$, $8 - 14 + 6 = 0$, $-27 + 21 + 6 = 0$), and their product/sum check against $x^3 - 7x + 6$. Here $p = -7$, $q = 6$, so

$$\Delta_0 = \frac{q^2}{4} + \frac{p^3}{27} = 9 - \frac{343}{27} = -\frac{100}{27} < 0,$$

the casus irreducibilis. Cardano's formula asserts (choosing conjugate cube roots so the sum is real)

$$1 = \sqrt[3]{-3 + \tfrac{10}{9}\sqrt{-3}} + \sqrt[3]{-3 - \tfrac{10}{9}\sqrt{-3}},$$

since $-q/2 = -3$ and $\sqrt{\Delta_0} = \sqrt{-100/27} = \frac{10}{9}\sqrt{-3}$. A visible integer named by two non-real cube roots — your own Bombelli identity.

Roots $1, 2, -3$; $\Delta_0 = -\frac{100}{27}$; $1 = \sqrt[3]{-3 + \frac{10}{9}\sqrt{-3}} + \sqrt[3]{-3 - \frac{10}{9}\sqrt{-3}}$.

Problem 11. Show a rational root forces the formula to denest.

SOLUTION. Suppose $t^3 + pt + q$ ($p, q \in \mathbb{Q}$) has a rational root r. Factor

$$t^3 + pt + q = (t - r)(t^2 + rt + \beta), \qquad \beta = p + r^2,$$

(the linear-coefficient match gives $\beta - r^2 = p$, and $-r\beta = q$ holds because r is a root). Write the other two roots as s, s' with $s + s' = -r$, $ss' = \beta$. The cubic's discriminant $D = -4p^3 - 27q^2 = -108\,\Delta_0$ also equals $(r-s)^2(r-s')^2(s-s')^2$. Now

$$(r-s)(r-s') = r^2 + r^2 + \beta = 3r^2 + p, \qquad (s-s')^2 = r^2 - 4\beta = -3r^2 - 4p,$$

so $D = (3r^2 + p)^2(-3r^2 - 4p)$. Hence

$$\Delta_0 = -\tfrac{1}{108}D = \tfrac{1}{108}(3r^2 + p)^2\,(3r^2 + 4p),$$

and $\sqrt{\Delta_0}$ is a rational multiple of $\sqrt{3(3r^2 + 4p)}$ — it lies in the *single* quadratic field $\mathbb{Q}(\sqrt{d})$, $d = 3(3r^2 + 4p)$. Therefore $u^3 = -\frac{q}{2} + \sqrt{\Delta_0} \in \mathbb{Q}(\sqrt{d})$, and expanding $u = m + n\sqrt{d}$ and matching parts shows such rational m, n exist: the cube root denests. Thus a rational root is exactly what makes Cardano's radicals collapse to the form $m + n\sqrt{d}$.

$\sqrt{\Delta_0} \in \mathbb{Q}(\sqrt{3(3r^2 + 4p)})$, **a single quadratic field, so the cube root denests.**

Problem 12. $\cos 20°$, the casus irreducibilis, and trisection.

SOLUTION. Put $\theta = 20°$ in $\cos 3\theta = 4\cos^3\theta - 3\cos\theta$. Since $\cos 60° = \frac{1}{2}$, with $c = \cos 20°$,

$$4c^3 - 3c = \tfrac{1}{2} \quad \Longrightarrow \quad 8c^3 - 6c - 1 = 0.$$

This has no rational root ($\pm 1, \pm\frac{1}{2}, \pm\frac{1}{4}, \pm\frac{1}{8}$ all fail), so it is irreducible over $\mathbb{Q}$. Normalizing to $c^3 - \frac{3}{4}c - \frac{1}{8} = 0$ gives $p = -\frac{3}{4}$, $q = -\frac{1}{8}$ and

$$\Delta_0 = \frac{q^2}{4} + \frac{p^3}{27} = \frac{1}{256} - \frac{1}{64} = -\frac{3}{256} < 0,$$

the casus irreducibilis (its three real roots are $\cos 20°, \cos 140°, \cos 260°$). So $\cos 20°$ cannot be written with real radicals. **Trisection.** A length is constructible by straightedge and compass exactly when it lies in a tower of *real quadratic* extensions of $\mathbb{Q}$ — so every constructible number has degree a power of 2 over $\mathbb{Q}$. But $\cos 20°$ has degree 3 (its minimal polynomial $8c^3 - 6c - 1$ is irreducible), and 3 is not a power of 2; hence $\cos 20°$ is not constructible. Constructing it is exactly what trisecting the constructible angle $60°$ would require, so $60°$ cannot be trisected with straightedge and compass.

$\cos 20°$ **has degree** 3**, hence is non-constructible, so** $60°$ **can't be trisected.**

Tools From the Imaginary Box

Full worked solutions. Each one trades geometry for algebra (or back), using "multiply lengths, add angles."

Launch

Problem 1. Compute $(1+i)^2$ and $(1+i)^4$ as length and angle.

SOLUTION. Directly, $(1+i)^2 = 1+2i+i^2 = 2i$. In polar terms $1+i$ has length $\sqrt{2}$ and angle $45°$; squaring squares the length to 2 and doubles the angle to $90°$, giving $2i$. Squaring again, $(1+i)^4 = (2i)^2 = -4$ — length 4, angle $180°$.

$$(1+i)^2 = 2i, \ (1+i)^4 = -4$$

Problem 2. Find the four solutions of $z^4 = 1$.

SOLUTION. The fourth roots of unity are arrows of length 1 at angles $0°, 90°, 180°, 270°$: that is $z = 1, i, -1, -i$, the corners of a square inscribed in the unit circle.

$$1, i, -1, -i$$

Core

Problem 3. Derive $\cos 3\theta = 4\cos^3\theta - 3\cos\theta$ via de Moivre.

SOLUTION. Expand the cube, take the real part. By de Moivre, $\cos 3\theta + i\sin 3\theta = (\cos\theta + i\sin\theta)^3$. Expand the right side with the binomial theorem:

$$\cos^3\theta + 3\cos^2\theta(i\sin\theta) + 3\cos\theta(i\sin\theta)^2 + (i\sin\theta)^3.$$

The real part (terms with even powers of i) is $\cos^3\theta - 3\cos\theta\sin^2\theta$. Substitute $\sin^2\theta = 1-\cos^2\theta$:

$$\cos 3\theta = \cos^3\theta - 3\cos\theta(1-\cos^2\theta) = 4\cos^3\theta - 3\cos\theta.$$

■

Problem 4. Prove the sum of all nth roots of unity is 0 for $n \geq 2$.

SOLUTION. Multiply the sum by ζ to rotate. Let $\zeta = e^{2\pi i/n}$, so the roots are $1, \zeta, \zeta^2, \dots, \zeta^{n-1}$, and let S be their sum. Multiplying by ζ sends each root to the next one (and ζ^{n-1} back to $\zeta^n = 1$), merely permuting the roots, so $\zeta S = S$. Then $(\zeta - 1)S = 0$. Since $n \geq 2$ means $\zeta \neq 1$, we conclude $S = 0$. ■

Problem 5. Compute $\binom{n}{0} + \binom{n}{2} + \binom{n}{4} + \cdots$ via the filter with $\omega = -1$.

SOLUTION. Average $(1+x)^n$ at $x = 1$ and $x = -1$. Adding $(1+1)^n$ and $(1-1)^n$ kills the odd-index terms and doubles the even-index ones:

$$(1+1)^n + (1-1)^n = \sum_k \binom{n}{k}(1+(-1)^k) = 2\sum_{k \text{ even}} \binom{n}{k}.$$

So $\sum_{k \text{ even}} \binom{n}{k} = \frac{2^n + 0^n}{2} = 2^{n-1}$ (for $n \geq 1$). **The even-index row sum is 2^{n-1}.**

2^{n-1}

Stretch

Problem 6. Find $\binom{n}{0} + \binom{n}{3} + \binom{n}{6} + \cdots$ via the cube-root filter; evaluate at $n = 9$.

SOLUTION. Average over the three cube roots of unity. Let $\omega = e^{2\pi i/3}$. Evaluating $(1+x)^n$ at $x = 1, \omega, \omega^2$ and adding, the coefficient of $\binom{n}{k}$ becomes $1 + \omega^k + \omega^{2k}$, which is 3 when $3 \mid k$ and 0 otherwise. Hence

$$\binom{n}{0} + \binom{n}{3} + \cdots = \frac{(1+1)^n + (1+\omega)^n + (1+\omega^2)^n}{3} = \frac{2^n + 2\cos\frac{n\pi}{3}}{3}.$$

For $n = 9$: $\cos 3\pi = -1$, so the sum is $\frac{512+2(-1)}{3} = \frac{510}{3} = 170$.

$\frac{2^n + 2\cos(n\pi/3)}{3}$; 170 **for** $n = 9$

Problem 7. Prove $\prod_{k=1}^{n-1} 2\sin\frac{k\pi}{n} = n$ by factoring $z^n - 1$; verify for $n = 4$.

SOLUTION. Factor $z^n - 1$ by its roots, then set $z = 1$. The roots of $z^n - 1$ are ζ^k, $k = 0, \dots, n-1$, with $\zeta = e^{2\pi i/n}$. Pulling out the $k = 0$ factor $(z-1)$,

$$\prod_{k=1}^{n-1}(z - \zeta^k) = \frac{z^n - 1}{z - 1} = 1 + z + \cdots + z^{n-1}.$$

Setting $z = 1$ makes the right side n, so $\prod_{k=1}^{n-1}(1-\zeta^k) = n$. Each factor is a chord of the unit circle from 1 to ζ^k, of length $|1-\zeta^k| = 2\sin\frac{k\pi}{n}$ (isosceles triangle, two unit radii at central angle $\frac{2\pi k}{n}$). Taking absolute values, $\prod_{k=1}^{n-1} 2\sin\frac{k\pi}{n} = n$.

Check $n = 4$. $2\sin 45° \cdot 2\sin 90° \cdot 2\sin 135° = \sqrt{2}\cdot 2\cdot\sqrt{2} = 4 = n$. ■

$\prod 2\sin\frac{k\pi}{n} = n;\ = 4$ **for** $n = 4$

Problem 8. Derive $\cos 36° = \frac{1+\sqrt{5}}{4}$ from the fifth roots of unity; how does φ appear?

SOLUTION. The real parts of the fifth roots sum to zero. The five fifth-roots of unity sum to 0; their real parts give $1+2\cos 72°+2\cos 144° = 0$. Since $\cos 144° = -\cos 36°$, this becomes $1+2\cos 72°-2\cos 36° = 0$, i.e. $\cos 36° - \cos 72° = \frac{1}{2}$.

Solve the quadratic. With $c = \cos 36°$ and $\cos 72° = 2c^2 - 1$ (double angle), $c - (2c^2-1) = \frac{1}{2}$, so $4c^2 - 2c - 1 = 0$ and (taking the positive root)

$$c = \frac{2+\sqrt{20}}{8} = \frac{1+\sqrt{5}}{4} \approx 0.809.$$

This is $\frac{\varphi}{2}$ with $\varphi = \frac{1+\sqrt{5}}{2}$ the golden ratio: $\sqrt{5}$ enters because the pentagon's diagonal-to-side ratio is φ, and the pentagon's geometry *is* the fifth roots of unity. ■

$\cos 36° = \frac{1+\sqrt{5}}{4} = \frac{\varphi}{2}$

The Flash

Problem 9. What is i^{2026}?

SOLUTION. Powers of i cycle with period 4. Since $2026 \equiv 2 \pmod 4$, $i^{2026} = i^2 = -1$.

-1

Problem 10. Evaluate $\cos 0 + \cos\frac{2\pi}{5} + \cos\frac{4\pi}{5} + \cos\frac{6\pi}{5} + \cos\frac{8\pi}{5}$.

SOLUTION. These five cosines are the real parts of the five fifth-roots of unity, whose sum is 0. The real part of 0 is 0, so the cosine sum is 0.

0

Proof Lab

Problem 11. Prove de Moivre's theorem by induction.

SOLUTION. Induct on n. Claim: $(\cos\theta + i\sin\theta)^n = \cos n\theta + i\sin n\theta$. The base case $n = 1$ is an identity. Assume it for n; then

$$(\cos\theta + i\sin\theta)^{n+1} = (\cos n\theta + i\sin n\theta)(\cos\theta + i\sin\theta).$$

Multiply out and use the angle-addition formulas: the real part is $\cos n\theta\cos\theta - \sin n\theta\sin\theta = \cos((n+1)\theta)$, and the imaginary part is $\sin n\theta\cos\theta + \cos n\theta\sin\theta = \sin((n+1)\theta)$. So the claim holds for $n+1$. (Geometrically: multiplying an angle-$n\theta$ unit arrow by an angle-θ unit arrow adds the angles to $(n+1)\theta$.) ■

Problem 12. Prove the roots-of-unity filter.

SOLUTION. The key sum $\sum_j \omega^{jm}$ is a switch. Let $\omega = e^{2\pi i/d}$ be a primitive dth root of unity, and let $f(x) = \sum_m c_m x^m$. Consider the average

$$\frac{1}{d}\sum_{j=0}^{d-1} f(\omega^j x) = \frac{1}{d}\sum_{j=0}^{d-1}\sum_m c_m\omega^{jm}x^m = \sum_m c_m x^m\left(\frac{1}{d}\sum_{j=0}^{d-1}\omega^{jm}\right).$$

The inner sum $\frac{1}{d}\sum_{j=0}^{d-1}\omega^{jm}$ is a geometric series in ω^m. If $d \mid m$ then $\omega^m = 1$ and every term is 1, giving average 1. If $d \nmid m$ then $\omega^m \neq 1$ and the full geometric sum over a complete set of roots vanishes, giving 0. So the average keeps exactly the terms $c_m x^m$ with $d \mid m$ and discards the rest — precisely the claimed filter. ■

Problem 13. Show that exactly 24 subsets of $\{1, \ldots, 6\}$ have sum divisible by 3.

SOLUTION. Pack subsets into $P(x) = \prod(1 + x^j)$ and filter with cube roots. The coefficient of x^s in $P(x) = \prod_{j=1}^{6}(1 + x^j)$ counts subsets with sum s, so the number with sum divisible by 3 is $\frac{1}{3}(P(1) + P(\omega) + P(\omega^2))$ for $\omega = e^{2\pi i/3}$. Now $P(1) = 2^6 = 64$. Grouping the factors of $P(\omega)$ by j mod 3,

$$P(\omega) = (1+\omega)^2(1+\omega^2)^2(1+1)^2 = \left[(1+\omega)(1+\omega^2)\right]^2 \cdot 4 = 1^2 \cdot 4 = 4,$$

using $(1+\omega)(1+\omega^2) = 1 + \omega + \omega^2 + \omega^3 = 1$; likewise $P(\omega^2) = 4$. Hence the count is $\frac{64+4+4}{3} = 24$. ■

24

Problem 14. Prove the Brahmagupta–Fibonacci identity via $|zw| = |z|\,|w|$, and write 65 as a sum of two squares.

SOLUTION. A sum of two squares is a norm. Let $z = a + bi$ and $w = c + di$, so $|z|^2 = a^2 + b^2$ and $|w|^2 = c^2 + d^2$. Since $|zw| = |z|\,|w|$, squaring gives $|zw|^2 = (a^2+b^2)(c^2+d^2)$. But $zw = (ac-bd) + (ad+bc)i$, so $|zw|^2 = (ac-bd)^2 + (ad+bc)^2$. Equating,

$$(a^2+b^2)(c^2+d^2) = (ac-bd)^2 + (ad+bc)^2.$$

For $65 = 5 \cdot 13 = (1^2 + 2^2)(2^2 + 3^2)$, take $z = 1 + 2i$, $w = 2 + 3i$: then $zw = -4 + 7i$, giving $65 = 4^2 + 7^2$. (Using $\bar{w} = 2 - 3i$ gives $zw = 8 + i$, so also $65 = 8^2 + 1^2$.) ■

$$65 = 4^2 + 7^2 = 8^2 + 1^2$$

Problem 15. Prove that a prime $p \equiv 3 \pmod 4$ is never a sum of two squares.

SOLUTION. Reduce mod 4. Every square is 0 or 1 $\pmod 4$, so a sum of two squares is $\equiv 0, 1$, or 2 $\pmod 4$ — never 3. A prime $p \equiv 3 \pmod 4$ therefore cannot be written as $a^2 + b^2$. (By contrast, a prime $p \equiv 1 \pmod 4$ splits in $\mathbb{Z}[i]$ as $(a + bi)(a - bi) = a^2 + b^2$, so those primes always are sums of two squares — Fermat's two-square theorem.) ■

The Picture Behind the Formula

Full worked solutions. Each one reads a tangle of square roots as a picture — a path, a difference of distances, a squashed circle — and lets the geometry hand over the answer.

Launch

Problem 1. Minimize $\sqrt{x^2+1}+\sqrt{(4-x)^2+4}$ over real x.

SOLUTION. Each root is a distance. With $P=(x,0)$ on the x-axis, $\sqrt{x^2+1}=PA$ for $A=(0,1)$ and $\sqrt{(4-x)^2+4}=PB$ for $B=(4,2)$. Both fixed points sit above the axis, so reflect B to $B'=(4,-2)$; then $PB=PB'$ and the sum is $PA+PB'$, minimized by the straight segment AB':

$$\min = AB' = \sqrt{(4-0)^2+(-2-1)^2} = \sqrt{16+9} = 5.$$

5

Problem 2. A circle has radius 5; a point P is 13 from its center. Length of the tangent from P?

SOLUTION. Tangent, radius, and center form a right triangle. The tangent is perpendicular to the radius at the point of contact, so $PT^2=d^2-r^2=13^2-5^2=144$, giving $PT=12$. (This is the power of the point, $\sqrt{d^2-r^2}$.)

12

Core

Problem 3. Minimize $\sqrt{(x-1)^2+4}+\sqrt{(x-5)^2+9}$.

SOLUTION. Read as a path, then reflect. With $P = (x, 0)$, the terms are PA and PB for $A = (1, 2)$, $B = (5, 3)$. Reflect B across the x-axis to $B' = (5, -3)$; the minimum is the straight distance

$$AB' = \sqrt{(5-1)^2 + (-3-2)^2} = \sqrt{16+25} = \sqrt{41}. \blacksquare$$

$\sqrt{41}$

Problem 4. Two chords cross; one is split into 3 and 8, the other into 4 and x. Find x.

SOLUTION. Intersecting chords: equal products. $PA \cdot PB = PC \cdot PD$ gives $3 \cdot 8 = 4 \cdot x$, so $x = 6$.

$x = 6$

Problem 5. Minimize $\sqrt{x^2 + y^2}$ subject to $3x + 4y = 25$.

SOLUTION. A constrained root is a distance to a line. $\sqrt{x^2 + y^2}$ is the distance from the origin to (x, y), and the constraint pins (x, y) to a line; the smallest such distance is the perpendicular distance,

$$\frac{|3 \cdot 0 + 4 \cdot 0 - 25|}{\sqrt{3^2 + 4^2}} = \frac{25}{5} = 5,$$

attained at the foot $(3, 4)$. $\blacksquare$

5

Stretch

Problem 6. Maximize $\sqrt{(x+1)^2 + 9} - \sqrt{(x-3)^2 + 1}$.

SOLUTION. A difference of distances, capped by the reverse triangle inequality. With $P = (x, 0)$, $A = (-1, 3)$, $B = (3, 1)$, the expression is $PA - PB \leq AB$, with equality when P, B, A are collinear and B lies between P and A. Here

$$AB = \sqrt{(3+1)^2 + (1-3)^2} = \sqrt{16+4} = 2\sqrt{5}.$$

Line AB ($y = 2.5 - \frac{x}{2}$) meets the axis at $x = 5$, where $P = (5, 0)$ lies beyond B, so the bound is achieved: the maximum is $2\sqrt{5}$ at $x = 5$. $\blacksquare$

$2\sqrt{5}$, **at** $x = 5$

Problem 7. Minimize $\sqrt{x^2 + 4} + \sqrt{(6-x)^2 + 9}$ with Minkowski's inequality.

SOLUTION. Two vectors, tip to tail. Let $\vec{u} = (x, 2)$ and $\vec{v} = (6 - x, 3)$, so the sum is $|\vec{u}| + |\vec{v}|$ and $\vec{u} + \vec{v} = (6, 5)$ regardless of x. By the triangle inequality,

$$|\vec{u}| + |\vec{v}| \geq |\vec{u} + \vec{v}| = \sqrt{6^2 + 5^2} = \sqrt{61},$$

with equality when $\vec{u} \parallel \vec{v}$ (same direction): $\frac{x}{2} = \frac{6-x}{3}$, i.e. $x = 2.4$. The minimum is $\sqrt{61}$. ■

$\sqrt{61}$, **at** $x = 2.4$

Problem 8. From an external point, a tangent has length 6 and a secant meets the circle at distances 4 and x. Find x.

SOLUTION. Tangent-squared equals the secant product. $PT^2 = PA \cdot PB$, so $6^2 = 4x$, giving $x = 9$.

$x = 9$

Problem 9. Find the tangent to $\frac{x^2}{16} + \frac{y^2}{9} = 1$ at (x_0, y_0) on it, by squashing to a circle.

SOLUTION. Squash, use the circle tangent, stretch back. The map $u = x/4$, $v = y/3$ sends the ellipse to the unit circle $u^2 + v^2 = 1$, and (x_0, y_0) to $(u_0, v_0) = (x_0/4,\ y_0/3)$. On the circle the tangent at (u_0, v_0) is $uu_0 + vv_0 = 1$. Substituting back,

$$\frac{x}{4} \cdot \frac{x_0}{4} + \frac{y}{3} \cdot \frac{y_0}{3} = 1 \quad \Longrightarrow \quad \frac{xx_0}{16} + \frac{yy_0}{9} = 1. ■$$

$$\frac{xx_0}{16} + \frac{yy_0}{9} = 1$$

Problem 10. A triangle is inscribed in $\frac{x^2}{25} + \frac{y^2}{9} = 1$. Find its maximum area.

SOLUTION. Borrow the circle's champion, then rescale. Squashing by $u = x/5$, $v = y/3$ turns the ellipse into the unit circle and takes inscribed triangles to inscribed triangles, preserving area *ratios*. The largest triangle inscribed in a unit circle is equilateral, area $\frac{3\sqrt{3}}{4}$. Since areas scale by $ab = 5 \cdot 3 = 15$ under the inverse stretch, the maximum inscribed area on the ellipse is

$$\frac{3\sqrt{3}}{4} \cdot 15 = \frac{45\sqrt{3}}{4}. ■$$

$$\frac{45\sqrt{3}}{4}$$

The Flash

Problem 11. Minimize $\sqrt{x^2+4}+\sqrt{(3-x)^2+4}$.

SOLUTION. With $P=(x,0)$, $A=(0,2)$, $B=(3,2)$; reflect $B\to(3,-2)$, so the minimum is $AB'=\sqrt{3^2+4^2}=5$ (at $x=1.5$).

5

Problem 12. Secant from P meets the circle at A then B with $PA=4$, $AB=5$; tangent touches at T. Find PT.

SOLUTION. The far intersection is $PB=PA+AB=9$, so $PT^2=PA\cdot PB=4\cdot 9=36$ and $PT=6$.

6

Proof Lab

Problem 13. Prove the reflection principle: for A,B on the same side of a line ℓ, the minimizer of $AP+PB$ over $P\in\ell$ is $AB'\cap\ell$, where B' is the reflection of B across ℓ.

SOLUTION. Reflection preserves the distance to ℓ. Since ℓ is the perpendicular bisector of BB', every point $P\in\ell$ satisfies $PB=PB'$. Hence for all $P\in\ell$,

$$AP+PB=AP+PB' \;\geq\; AB',$$

by the triangle inequality applied to the path $A\to P\to B'$. Equality holds exactly when P lies on the straight segment AB'. Because A and B' are on *opposite* sides of ℓ (we reflected B across it), the segment AB' does cross ℓ, at a unique point $P^*=AB'\cap\ell$. That P^* achieves the bound AB', so it is the minimizer.
■

Problem 14. Prove $\sqrt{a^2+b^2}+\sqrt{c^2+d^2}\geq\sqrt{(a+c)^2+(b+d)^2}$, with the equality condition.

SOLUTION. Triangle inequality for vectors. Let $\vec{u}=(a,b)$ and $\vec{v}=(c,d)$. The claim is $|\vec{u}|+|\vec{v}|\geq|\vec{u}+\vec{v}|$. Square the (nonnegative) left side and use $\vec{u}\cdot\vec{v}\leq|\vec{u}|\,|\vec{v}|$ (Cauchy–Schwarz):

$$(|\vec{u}|+|\vec{v}|)^2=|\vec{u}|^2+2|\vec{u}||\vec{v}|+|\vec{v}|^2\;\geq\;|\vec{u}|^2+2(\vec{u}\cdot\vec{v})+|\vec{v}|^2=|\vec{u}+\vec{v}|^2.$$

Taking square roots gives the inequality. Equality needs $\vec{u}\cdot\vec{v}=|\vec{u}||\vec{v}|$, i.e. $\vec{u}$ and $\vec{v}$ are *parallel and point the same way* (one is a nonnegative scalar multiple of the other). ■

Problem 15. Prove the power of a point P equals d^2-r^2 for every line through P, and explain the tangent limit.

SOLUTION. Anchor with the secant through the center. Let P be at distance d from the center of a circle of radius r. The secant through P *and the center* meets the circle at distances $d - r$ and $d + r$ from P, whose product is

$$(d - r)(d + r) = d^2 - r^2.$$

Every secant gives the same product. For any other secant through P meeting the circle at A, B, the intersecting-chords / tangent-secant similar triangles show $PA \cdot PB$ is the same constant — so it equals $d^2 - r^2$ for *every* line through P (with directed lengths carrying the sign; an interior point gives the negative value).

The tangent as a limit. Rotate a secant until its two intersection points A and B slide together into a single point T: the line becomes tangent, and $PA \cdot PB \to PT \cdot PT = PT^2$. Hence $PT^2 = d^2 - r^2$, the tangent case as the limiting secant. ■

The Lever and the Shadow

Full worked solutions. Each one either hangs weights on the corners and reads a ratio off the see-saw, slices the figure into areas, or flattens it into a convenient shadow and finishes with coordinates.

Launch

Problem 1. Masses 2 at A and 7 at B on segment AB: how does the balance point divide AB?

SOLUTION. The balance point hugs the heavier mass. By the lever law $AP : PB = m_B : m_A$, the balance point divides AB as $7 : 2$ — closer to B, the heavier end.

$$AP : PB = 7 : 2$$

Problem 2. D on BC with $BD : DC = 3 : 4$: what masses at B, C make D the balance point?

SOLUTION. Masses are inverse to the segments. A balance point satisfies $BD : DC = m_C : m_B$, so $m_C : m_B = 3 : 4$, i.e. $m_B : m_C = 4 : 3$. The smallest integers are $m_B = 4$, $m_C = 3$ (and then D carries mass 7).

$$m_B = 4, \ m_C = 3$$

Core

Problem 3. D midpoint of BC, E on AC with $AE : EC = 1 : 2$; $AD \cap BE = P$. Find $AP : PD$, $BP : PE$.

SOLUTION. Set one mass, let the rest cascade. D is the midpoint, so $m_B = m_C$; take $m_B = m_C = 1$. For E on AC with $AE : EC = 1 : 2$ we need $m_A : m_C = EC : AE = 2 : 1$, so $m_A = 2$. The balance points carry

$m_D = m_B + m_C = 2$ and $m_E = m_A + m_C = 3$. Reading each cevian as a see-saw,

$$AP : PD = m_D : m_A = 2 : 2 = 1 : 1, \qquad BP : PE = m_E : m_B = 3 : 1. \blacksquare$$

$$AP : PD = 1 : 1, \quad BP : PE = 3 : 1$$

Problem 4. $BD : DC = 2 : 1, CE : EA = 1 : 2$; for what $AF : FB$ are AD, BE, CF concurrent?

SOLUTION. Ceva's product must equal 1. Concurrency requires

$$\frac{BD}{DC} \cdot \frac{CE}{EA} \cdot \frac{AF}{FB} = 1 \quad \Longrightarrow \quad \frac{2}{1} \cdot \frac{1}{2} \cdot \frac{AF}{FB} = 1 \quad \Longrightarrow \quad \frac{AF}{FB} = 1.$$

So F must be the midpoint of AB. ■

$$AF : FB = 1 : 1$$

Problem 5. Using masses $1, 1, 1$, show the medians meet dividing each $2 : 1$ from the vertex.

SOLUTION. Equal masses balance everywhere. Put mass 1 at each vertex. Every side is split $1 : 1$, so each midpoint is a balance point carrying mass 2, and all three medians pass through the common center of mass $G = \frac{1}{3}(A + B + C)$ (the centroid). Along a median, G balances a vertex (mass 1) against the opposite midpoint (mass 2), so

$$(\text{vertex})\, G : G\, (\text{midpoint}) = m_{\text{mid}} : m_{\text{vertex}} = 2 : 1. \blacksquare$$

Each median is divided $2 : 1$ from the vertex; the medians are concurrent at the centroid.

Stretch

Problem 6. $BD : DC = 1 : 2, AF : FB = 1 : 3; AD \cap CF = P$. Find $AP : PD$ and $CP : PF$.

SOLUTION. Weigh three corners. From $BD : DC = 1 : 2$ we need $m_B : m_C = 2 : 1$; take $m_B = 2$, $m_C = 1$. From $AF : FB = 1 : 3$ we need $m_A : m_B = FB : AF = 3 : 1$, so $m_A = 3m_B = 6$. The balance points carry $m_D = m_B + m_C = 3$ and $m_F = m_A + m_B = 8$. Hence

$$AP : PD = m_D : m_A = 3 : 6 = 1 : 2, \qquad CP : PF = m_F : m_C = 8 : 1. \blacksquare$$

$$AP : PD = 1 : 2, \quad CP : PF = 8 : 1$$

Problem 7. Sides split $2:1$ in the same sense ($BD:DC = CE:EA = AF:FB = 2:1$): area of the central triangle?

SOLUTION. The cevians are *not* concurrent, so weights stall — cast a shadow. The Ceva product is $2\cdot2\cdot2 = 8 \neq 1$, so AD, BE, CF bound a central triangle. Since area-ratios are affine-invariant, normalize to a convenient right triangle with legs divisible by 3:

$$A = (0,0), \quad B = (3,0), \quad C = (0,3).$$

Then $F = (2,0)$ (as $AF:FB = 2:1$), $E = (0,1)$ (as $CE:EA = 2:1$, i.e. $AE:EC = 1:2$), and $D = B + \frac{2}{3}(C - B) = (1,2)$. The three cevians are

$$AD: \ y = 2x, \qquad BE: \ x + 3y = 3, \qquad CF: \ 3x + 2y = 6.$$

Their pairwise intersections are

$$S = AD \cap BE = \left(\tfrac{3}{7}, \tfrac{6}{7}\right), \quad T = BE \cap CF = \left(\tfrac{12}{7}, \tfrac{3}{7}\right), \quad R = CF \cap AD = \left(\tfrac{6}{7}, \tfrac{12}{7}\right).$$

The shoelace formula gives $[STR] = \frac{9}{14}$, while $[ABC] = \frac{1}{2}\cdot3\cdot3 = \frac{9}{2}$, so

$$\frac{[STR]}{[ABC]} = \frac{9/14}{9/2} = \frac{1}{7}.$$

The central triangle is one-seventh of the whole — the classic "one-seventh area" triangle. ∎

$\frac{1}{7}$ **of the triangle**

Problem 8. Parallelogram $ABCD$: E midpoint of BC, F on CD with $CF:FD = 1:2$; $AE \cap BF = P$. Find $AP:PE$.

SOLUTION. Every parallelogram is an affine square. Map $ABCD$ to $A = (0,0)$, $B = (1,0)$, $C = (1,1)$, $D = (0,1)$; then $E = (1, \frac{1}{2})$ and $F = (\frac{2}{3}, 1)$. Line AE is $y = \frac{1}{2}x$; line BF through $(1,0)$ and $(\frac{2}{3},1)$ is $y = -3x + 3$. Setting them equal, $\frac{1}{2}x = -3x + 3$ gives $x = \frac{6}{7}$, so $P = (\frac{6}{7}, \frac{3}{7})$, which is $\frac{6}{7}$ of the way from A to E:

$$AP:PE = 6:1. \blacksquare$$

$AP:PE = 6:1$

The Flash

Problem 9. Masses 5 at A, 3 at B: ratio in which AB is divided?

SOLUTION. $AP:PB = m_B:m_A = 3:5$.

$3:5$

Problem 10. $BD : DC = 3 : 4$, $AE : EC = 3 : 4$; $AD \cap BE = P$. Find $BP : PE$.

SOLUTION. From $BD : DC = 3 : 4$, $m_B : m_C = 4 : 3$; take $m_B = 4$, $m_C = 3$. From $AE : EC = 3 : 4$, $m_A : m_C = EC : AE = 4 : 3$, so $m_A = 4$. Then $m_E = m_A + m_C = 7$, and $BP : PE = m_E : m_B = 7 : 4$.

$$BP : PE = 7 : 4$$

Proof Lab

Problem 11. Prove the see-saw law: m_A at A, m_B at B balance at P with $AP : PB = m_B : m_A$.

SOLUTION. The balance point is the center of mass. The center of mass of the two weighted points is

$$P = \frac{m_A A + m_B B}{m_A + m_B}.$$

Rewrite it relative to A: $P = A + \frac{m_B}{m_A + m_B}(B - A)$, so P lies on segment AB at parameter $t = \frac{m_B}{m_A + m_B}$ from A. Therefore

$$AP : PB = t : (1 - t) = \frac{m_B}{m_A + m_B} : \frac{m_A}{m_A + m_B} = m_B : m_A.$$

Equivalently, the torque condition $m_A \cdot AP = m_B \cdot PB$ rearranges directly to $AP : PB = m_B : m_A$. Either way, the balance point sits closer to the heavier mass. ∎

Problem 12. Prove Ceva's theorem by mass points.

SOLUTION. Force two cevians through the center of mass, then test the third. Assign vertex masses m_A, m_B, m_C so that D and E are balance points:

$$\frac{BD}{DC} = \frac{m_C}{m_B}, \qquad \frac{CE}{EA} = \frac{m_A}{m_C}.$$

(One can always solve for such masses up to scale.) Let $G = \frac{m_A A + m_B B + m_C C}{m_A + m_B + m_C}$ be the center of mass. Because G is the balance point of the whole system, it lies on the segment from each vertex to the balance point of the other two — so G lies on both AD and BE; the two cevians meet at G.

Now the third cevian CF passes through G if and only if F is the balance point of A and B, that is $\frac{AF}{FB} = \frac{m_B}{m_A}$. Multiplying the three side-ratios,

$$\frac{BD}{DC} \cdot \frac{CE}{EA} \cdot \frac{AF}{FB} = \frac{m_C}{m_B} \cdot \frac{m_A}{m_C} \cdot \frac{AF}{FB} = \frac{m_A}{m_B} \cdot \frac{AF}{FB}.$$

This equals 1 exactly when $\frac{AF}{FB} = \frac{m_B}{m_A}$ — exactly when F is the balance point and CF passes through G. Hence the three cevians are concurrent if and only if the product is 1. ∎

Problem 13. Prove an affine map preserves division ratios and area ratios.

SOLUTION. Affinity respects barycentric averages and scales all areas equally. Let $f(\mathbf{x}) = M\mathbf{x} + \mathbf{b}$ with M invertible.

(*a*) *Division ratios.* A point dividing PQ in ratio $t : (1-t)$ is $X = (1-t)P + tQ$, a barycentric blend with coefficients summing to 1. Then

$$f(X) = M((1-t)P + tQ) + \mathbf{b} = (1-t)(MP + \mathbf{b}) + t(MQ + \mathbf{b}) = (1-t)f(P) + tf(Q),$$

using $(1-t) + t = 1$ to distribute the shift $\mathbf{b}$. So $f(X)$ divides $f(P)f(Q)$ in the same ratio $t : (1-t)$.

(*b*) *Area ratios.* A triangle with vertices P, Q, R has area $\frac{1}{2}|\det(Q-P,\ R-P)|$. Applying f, each edge vector $Q - P$ becomes $M(Q-P)$ (the shift $\mathbf{b}$ cancels in the difference), so

$$[f(P)f(Q)f(R)] = \tfrac{1}{2}|\det(M(Q-P),\ M(R-P))| = |\det M| \cdot [PQR].$$

Every area is multiplied by the same constant $|\det M|$, so the ratio of any two areas is unchanged. Together, (a) and (b) are exactly what license the normalization: solve in a convenient triangle, and both the division ratios and the area fractions transfer back. ■

Problem 14. $BD : DC = 1 : 3$, $CE : EA = 1 : 1$; find $[ABD]$ and $[CDE]$ (area $[ABC] = 1$).

SOLUTION. Same height, then shared vertex. Triangles ABD and ABC share the apex A with bases BD and BC on the same line, so by the same-height lemma

$$[ABD] = \frac{BD}{BC}[ABC] = \frac{1}{4}.$$

For $[CDE]$, the triangle CDE shares the vertex C with ABC, with D on CB (so $CD/CB = 3/4$) and E on CA (so $CE/CA = 1/2$). The shared-vertex lemma gives

$$[CDE] = \frac{CD}{CB} \cdot \frac{CE}{CA}[ABC] = \frac{3}{4} \cdot \frac{1}{2} = \frac{3}{8}. \blacksquare$$

$[ABD] = \frac{1}{4}, \quad [CDE] = \frac{3}{8}$

Problem 15. Sides each divided $1 : 2$ in the same sense; find $[DEF]/[ABC]$ by cutting corners.

SOLUTION. Each corner is a product of two side-fractions. The corner triangle at A is AFE, with F on AB ($AF/AB = 1/3$, since $AF : FB = 1 : 2$) and E on CA ($AE/AC = 2/3$, since $CE : EA = 1 : 2$). By the shared-vertex lemma,

$$\frac{[AFE]}{[ABC]} = \frac{AF}{AB} \cdot \frac{AE}{AC} = \frac{1}{3} \cdot \frac{2}{3} = \frac{2}{9}.$$

The rotational symmetry of the construction makes the corners at B and C identical, $\frac{2}{9}$ each. Removing all three,

$$\frac{[DEF]}{[ABC]} = 1 - 3 \cdot \frac{2}{9} = 1 - \frac{2}{3} = \frac{1}{3}. \blacksquare$$

$[DEF] = \frac{1}{3}[ABC]$

Problem 16. Cevians AD ($BD : DC = 1 : 2$) and CE ($AE : EB = 3 : 1$) meet at P; find $[APC]$ (area $[ABC] = 1$).

SOLUTION. Turn both side-splits into area ratios at P. Apply the cevian-to-a-point lemma to each cevian through P:

$$\frac{BD}{DC} = \frac{[ABP]}{[ACP]} = \frac{1}{2}, \qquad \frac{AE}{EB} = \frac{[ACP]}{[BCP]} = \frac{3}{1}.$$

Let $[BCP] = s$. The second ratio gives $[ACP] = 3s$; the first gives $[ABP] = \frac{1}{2}[ACP] = \frac{3}{2}s$. Since the three corner areas fill the triangle,

$$[ABP] + [BCP] + [ACP] = \tfrac{3}{2}s + s + 3s = \tfrac{11}{2}s = 1 \implies s = \frac{2}{11},$$

so $[APC] = [ACP] = 3s = \dfrac{6}{11}$. ■

$[APC] = \frac{6}{11}$

Counting Past the Mirror

Full worked solutions. Each one refuses to list-and-dedupe; instead it asks what every symmetry leaves fixed and averages the answers.

Launch

Problem 1. List the distinct 3-bead, 2-color necklaces under rotation, then confirm with Burnside.

SOLUTION. List, then check by averaging fixed-points. Under the rotation group C_3 the genuinely distinct necklaces are: all-black BBB, all-white WWW, "two black, one white" BBW, and "one black, two white" BWW — four in all (position stops mattering once you can rotate). Burnside confirms it: the identity fixes $2^3 = 8$ colorings, and each of the two nontrivial rotations is a single 3-cycle of the beads, so all three beads must match, fixing 2 each. Averaging,

$$\frac{2^3+2+2}{3} = \frac{12}{3} = 4. \blacksquare$$

4

Problem 2. How many 2-colorings does a 180° rotation of a 6-bead necklace fix?

SOLUTION. Cycle structure decides the fixed count. A 180° turn sends each bead to the one directly opposite, pairing the six positions into *three* 2-cycles. To look unchanged, the two beads in each pair must share a color, so there are $2^3 = 8$ fixed colorings (one free choice per cycle).

8

Core

Problem 3. Count the distinct 6-bead, 2-color necklaces under rotation (C_6) with Burnside.

SOLUTION. Rotation by k has $\gcd(k, 6)$ cycles. A rotation by k positions decomposes the 6 beads into $\gcd(k, 6)$ cycles, so it fixes $2^{\gcd(k,6)}$ colorings. Tabulating $k = 0, \ldots, 5$:

$$2^6,\ 2^1,\ 2^2,\ 2^3,\ 2^2,\ 2^1\ =\ 64,\ 2,\ 4,\ 8,\ 4,\ 2.$$

Averaging over the six rotations,

$$\frac{64+2+4+8+4+2}{6} = \frac{84}{6} = 14. \blacksquare$$

14

Problem 4. Count the distinct 4-bead, 2-color bracelets under the full dihedral group D_4 (flips allowed, so a coloring and its mirror image are identified), and compare with the rotation-only count of 6.

SOLUTION. Add the reflections to the rotations. The four rotations fix 16, 2, 4, 2 as before (sum 24). The four reflections split by axis: the two axes *through opposite beads* fix those two beads and swap the other pair — two fixed points and one 2-cycle, i.e. three cycles, fixing $2^3 = 8$ each; the two axes *through opposite edge-midpoints* swap the beads in two 2-cycles, fixing $2^2 = 4$ each. So the reflections contribute $8 + 8 + 4 + 4 = 24$. Averaging over all $|D_4| = 8$ symmetries,

$$\frac{24+24}{8} = \frac{48}{8} = 6.$$

Here it happens to equal the rotation-only count of 6; for this small bracelet the extra reflection-symmetry and the extra fixed-points balance out. ■

6 (**equal to the rotation-only count**)

Problem 5. How many ways to color the 4 corners of a square with 3 colors, under rotation?

SOLUTION. C_4 with 3 colors. The identity fixes $3^4 = 81$. The 90° and 270° rotations are each a single 4-cycle of the corners, so all four must match: 3 each. The 180° rotation is two 2-cycles, fixing $3^2 = 9$. Averaging,

$$\frac{81+3+9+3}{4} = \frac{96}{4} = 24. \blacksquare$$

24

Stretch

Problem 6. Count the distinct 3-colorings of the 6 faces of a cube under its 24 rotations.

SOLUTION. Group the 24 rotations by type and cycle-count the faces. With 3 colors, each rotation fixes $3^{(\#\text{face-cycles})}$:

- identity (1): six fixed faces → 3^6;
- 90°/270° face rotations (6): the four side faces form a 4-cycle, top and bottom fixed → 3 cycles → 3^3 each;
- 180° face rotations (3): the side faces form two 2-cycles, top/bottom fixed → 4 cycles → 3^4 each;
- 120°/240° vertex rotations (8): the faces split into two 3-cycles → 3^2 each;
- 180° edge rotations (6): the faces split into three 2-cycles → 3^3 each.

Averaging,

$$\frac{3^6 + 6 \cdot 3^3 + 3 \cdot 3^4 + 8 \cdot 3^2 + 6 \cdot 3^3}{24} = \frac{729 + 162 + 243 + 72 + 162}{24} = \frac{1368}{24} = 57. \blacksquare$$

57

Problem 7. Count the distinct 2-colorings of the edges of an equilateral triangle under D_3.

SOLUTION. The group acts on the three edges. D_3 has six elements. The identity fixes $2^3 = 8$. The two rotations $(120°, 240°)$ cycle all three edges, fixing 2 each. Each of the three reflections fixes one edge and swaps the other two — two cycles, fixing $2^2 = 4$ each. Averaging,

$$\frac{2^3 + 2 \cdot 2 + 3 \cdot 2^2}{6} = \frac{8 + 4 + 12}{6} = \frac{24}{6} = 4. \blacksquare$$

4

The Flash

Problem 8. How many distinct 2-colored cubes (faces) under rotation?

SOLUTION. Cycle-count the 24 rotations with 2 colors:

$$\frac{2^6 + 6 \cdot 2^3 + 3 \cdot 2^4 + 8 \cdot 2^2 + 6 \cdot 2^3}{24} = \frac{64 + 48 + 48 + 32 + 48}{24} = \frac{240}{24} = 10.$$

10

Problem 9. How many distinct 3-color corner-colorings of a triangle under rotation?

SOLUTION. C_3: the identity fixes $3^3 = 27$, and each nontrivial rotation fixes 3. So $\frac{27+3+3}{3} = \frac{33}{3} = 11$.

11

Proof Lab

Problem 10. Prove Burnside's lemma by counting fixed pairs (g, x) two ways.

SOLUTION. Double-count the incidences $g \cdot x = x$. Consider the set of pairs (g, x) with $g \in G$ and g fixing the arrangement x. Counting by g first,

$$\#\{(g,x) : g \cdot x = x\} = \sum_{g \in G} |\mathrm{Fix}(g)|.$$

Counting by x first, the same total is $\sum_x |\mathrm{Stab}(x)|$, where $\mathrm{Stab}(x) = \{g : g \cdot x = x\}$. By the orbit–stabilizer theorem $|\mathrm{Stab}(x)| = |G|/|\mathrm{orbit}(x)|$, so summing over a single orbit O,

$$\sum_{x \in O} |\mathrm{Stab}(x)| = |O| \cdot \frac{|G|}{|O|} = |G|.$$

Thus each orbit contributes exactly $|G|$, and summing over all orbits gives $\sum_x |\mathrm{Stab}(x)| = (\#\text{orbits}) \cdot |G|$. Equating the two counts,

$$\sum_{g \in G} |\mathrm{Fix}(g)| = (\#\text{orbits}) \cdot |G| \implies \#\text{orbits} = \frac{1}{|G|} \sum_{g \in G} |\mathrm{Fix}(g)|. \blacksquare$$

Problem 11. Derive the necklace formula $\frac{1}{n} \sum_{d|n} \varphi(d)\, c^{n/d}$.

SOLUTION. Sort the rotations by $\gcd(k, n)$. Under the rotation group C_n, a rotation by k positions decomposes the n beads into exactly $\gcd(k, n)$ cycles, so it fixes $c^{\gcd(k,n)}$ colorings. Burnside gives

$$\#\text{necklaces} = \frac{1}{n} \sum_{k=0}^{n-1} c^{\gcd(k,n)}.$$

Now group the terms by the value $d = \gcd(k, n)$, which must divide n. The number of $k \in \{0, \dots, n-1\}$ with $\gcd(k, n) = d$ is $\varphi(n/d)$ (write $k = dk'$ with $\gcd(k', n/d) = 1$). Therefore

$$\frac{1}{n} \sum_{d|n} \varphi\!\left(\frac{n}{d}\right) c^d = \frac{1}{n} \sum_{d|n} \varphi(d)\, c^{n/d},$$

the last step just reindexing $d \leftrightarrow n/d$ over the divisors of n. The Euler totient of **Chapter 10** is counting exactly the rotations of each cycle type. ■

Dots, Lines, and Lonely Hearts

Full worked solutions. Each one strips a problem down to vertices and edges, counts a local feature, and reads off a global verdict.

Launch

Problem 1. Can the "envelope" figure (a rectangle with both diagonals drawn in and a triangular flap on top) be drawn without lifting the pen? Count its odd-degree vertices.

SOLUTION. An Euler trail exists iff 0 or 2 vertices are odd. The envelope has five vertices: the roof apex (degree 2), the two upper rectangle corners (degree 4 each, where roof, side, and the two diagonals meet), and the two bottom corners (degree 3 each). Only the two bottom corners are odd. Two odd vertices means an Euler trail exists but not a closed circuit: start at one bottom corner and end at the other. Yes — **draw it in one stroke, starting and ending at the two bottom corners.**

Yes; 2 odd vertices (the bottom corners)

Problem 2. A graph has 6 vertices and 9 edges. What is the sum of all vertex degrees?

SOLUTION. Handshake lemma. Every edge contributes 2 to the degree sum, so $\sum \deg = 2E = 2 \cdot 9 = 18$.

Core

Problem 3. Prove that in any group of people, the number who shook an odd number of hands is even.

SOLUTION. The degree sum is even. Model the group as a graph: people are vertices, handshakes are edges. The handshake lemma gives $\sum_v \deg(v) = 2E$, an even number. Split the sum into even-degree and odd-degree vertices: the even-degree part is even, so the odd-degree part must also be even. But a sum of odd numbers is even only when there is an *even number* of them. Hence the count of odd-degree (odd-handshake) people is even. ■

Problem 4. Verify $V - E + F = 2$ for the octahedron ($V = 6, E = 12$); find F.

SOLUTION. Solve Euler's formula for F. Rearranging, $F = 2 - V + E = 2 - 6 + 12 = 8$. Indeed the octahedron has 8 triangular faces, and $6 - 12 + 8 = 2$. ■

$F = 8$

Problem 5. Show that K_5 cannot be drawn in the plane without crossings, using $E \leq 3V - 6$.

SOLUTION. Compare the edge count to the planar bound. Every simple connected planar graph with $V \geq 3$ satisfies $E \leq 3V - 6$ (a consequence of Euler's formula, since each face is bounded by at least three edges and each edge borders two faces). For K_5, $V = 5$ and $E = \binom{5}{2} = 10$, while $3V - 6 = 9$. Since $10 > 9$, the bound is violated, so K_5 cannot be planar. ■

Problem 6. Convert the $m \times n$ board into its grid graph. How many edges does it have? Check on the 2×3 board.

SOLUTION. Count horizontal and vertical adjacencies separately. In the grid graph, each cell is a vertex and side-adjacent cells are joined. Horizontal edges: each of the m rows has $n - 1$ consecutive pairs, giving $m(n-1)$. Vertical edges: each of the n columns has $m-1$ consecutive pairs, giving $n(m-1)$. Total

$$m(n - 1) + n(m - 1) = 2mn - m - n.$$

For the 2×3 board: $2 \cdot 6 - 2 - 3 = 12 - 5 = 7$ edges — indeed a 2×3 grid graph has 7 side-adjacencies. ■

$2mn - m - n$; 7 **for** 2×3

Problem 7. Soccer ball: 12 pentagons, 20 hexagons, three seams per corner. Find V, E, F and verify Euler.

SOLUTION. Two double counts, then Euler. Counting (face, bordering-seam) pairs, each pentagon contributes 5 and each hexagon 6, and each seam borders two faces: $5 \cdot 12 + 6 \cdot 20 = 180 = 2E$, so $E = 90$. Three seams meet at each corner and each seam has two ends: $3V = 2E = 180$, so $V = 60$. And $F = 12 + 20 = 32$. Check: $V - E + F = 60 - 90 + 32 = 2$. ■

$V = 60, E = 90, F = 32$

Problem 8. Draw the cube's graph in the plane and build its dual (a vertex in each face, joined when two faces share an edge). Which well-known graph results?

SOLUTION. The dual exchanges faces and vertices. The cube has 6 faces, and two faces are adjacent exactly when they share an edge — each face borders 4 others. So the dual has 6 vertices, each of degree 4, joined when the corresponding faces meet along an edge. That is precisely the *octahedron* graph. Cube and octahedron are dual Platonic solids, and Euler's counts confirm it: the cube's $(V, E, F) = (8, 12, 6)$ becomes the octahedron's $(6, 12, 8)$, with faces and vertices exchanged. **The octahedron graph.** ■

The octahedron graph

Problem 9. Decide whether $4, 4, 3, 1, 1, 1$ can be the degree sequence of a simple graph. Its sum is even — is that enough?

SOLUTION. Even sum is necessary, not sufficient; run Havel–Hakimi. The sum is $4 + 4 + 3 + 1 + 1 + 1 = 14$, even, so the handshake lemma is satisfied — but that is only the first filter. Apply the *Havel–Hakimi* reduction: sort largest-first, delete the top degree d, and subtract 1 from the next d entries. From $4, 4, 3, 1, 1, 1$: remove the leading 4 and subtract 1 from the next four, leaving $3, 2, 0, 0, 1$; re-sort to $3, 2, 1, 0, 0$. Now remove the leading 3 and subtract 1 from the next three entries $2, 1, 0$ — the last is a 0, which drops to -1. A negative degree is impossible, so the sequence is *not graphical*. (Directly: two degree-4 vertices among only six must each attach to four others, forcing each of the three degree-1 vertices above degree 1.) **No such simple graph exists.** ■

No — Havel–Hakimi yields a negative degree

Stretch

Problem 10. Prove that every tree with n vertices has exactly $n-1$ edges, by induction (remove a leaf).

SOLUTION. Induct on n, peeling a leaf. *Base case:* a tree with $n = 1$ vertex has $0 = 1 - 1$ edges. *Inductive step:* let T be a tree with $n \geq 2$ vertices. A finite tree has at least one *leaf* (a degree-1 vertex): follow a longest path and its endpoints must be leaves. Remove a leaf v and its single edge; the result is still connected and acyclic — a tree on $n-1$ vertices — which by the inductive hypothesis has $(n-1)-1 = n-2$ edges. Adding back v and its one edge gives $n-1$ edges for T. By induction, every tree on n vertices has $n-1$ edges. ■

Problem 11. Prove any 2-coloring of the edges of K_6 contains a monochromatic triangle.

SOLUTION. Pigeonhole at one vertex, then inspect the triangle. Fix a vertex P. It has 5 edges, each colored red or blue, so by pigeonhole at least $\lceil 5/2 \rceil = 3$ of them share a color — say P joins A, B, C all in red. Now look at the three edges among A, B, C. If *any* of them, say AB, is red, then P, A, B form a red triangle. If *none* is red, then A, B, C are pairwise blue — a blue triangle. Either way a monochromatic triangle is forced. ■

Problem 12. Workers w_1:$\{j_1, j_2\}$, w_2:$\{j_2, j_3\}$, w_3:$\{j_1, j_3\}$, w_4:$\{j_1, j_2, j_3\}$. Can all four get distinct jobs? Use Hall.

SOLUTION. Look for an over-subscribed group. Take $S = \{w_1, w_2, w_3, w_4\}$, all four workers. Their combined set of acceptable jobs is $N(S) = \{j_1, j_2, j_3\}$, since none of them will accept j_4. Then $|N(S)| = 3 < 4 = |S|$, which violates Hall's condition. So no perfect matching exists: four workers cannot be assigned to distinct jobs when together they accept only three. **No — the blocking group is all four workers.** ■

No; the blocking set is all four workers ($|N(S)| = 3 < 4$)

Problem 13. A beetle crawls through every cell of a 5 × 5 board once by side-steps. Prove the endpoints have the corner color; can it begin beside a corner?

SOLUTION. A cell-tour is a Hamiltonian path in a bipartite graph. Convert the board to its grid graph; the checkerboard coloring is its bipartition, and every side-step swaps color. A crawl visiting all 25 cells is a Hamiltonian path, so its cells alternate colors: with $25 = 13 + 12$, thirteen cells carry one color and twelve the other. In an alternating sequence of odd length, both endpoints share the majority color — the color of the four corners (each corner has coordinates summing to an even number, the 13-cell class). So both endpoints are corner-colored.

A cell *directly beside* a corner is one side-step away, hence the *minority* (12-cell) color. Since every crawl must start on a majority (corner-colored) cell, **no such crawl can begin beside a corner.** ■

Endpoints are corner-colored; a crawl cannot start beside a corner

Problem 14. At a party of n couples nobody shakes their own spouse; the host's $2n-1$ questions get $2n - 1$ different answers. How many hands did the host's spouse shake?

SOLUTION. Distinct answers force the full range; peel couples in pairs. Nobody shakes themselves or their spouse, so each of the $2n - 1$ other people shook between 0 and $2n - 2$ hands. Their $2n - 1$ answers are *distinct*, so they are exactly $0, 1, 2, \ldots, 2n - 2$, each once.

The person who answered $2n - 2$ shook everyone they possibly could — everyone except themselves and their own spouse. So every other person shook *that* person at least once; in particular the person who answered 0 shook nobody, and the only way both hold is that the $2n - 2$-shaker and the 0-shaker are *spouses*. Remove this couple. Everyone remaining had shaken the $(2n - 2)$-shaker (subtract 1 from their count) and had *not* shaken the 0-shaker (subtract nothing), so every remaining answer drops by exactly 1, and we face the identical situation with $n - 1$ couples. Iterating, the answerers pair off as spouses whose counts sum to $2n - 2$ (namely $\{0, 2n - 2\}, \{1, 2n - 3\}, \ldots$). The one person left without a partner among the answerers is the *host's own spouse*, holding the unpaired middle value.

host's spouse shook $n - 1$ hands.■

$n-1$

Problem 15. Let G be the friendship graph of the six people and $\overline{G}$ its complement (the stranger graph). Restate the six-people theorem in terms of triangles in G and $\overline{G}$.

SOLUTION. Friend/stranger is a graph and its complement. Put an edge of G between each friendly pair; then the complement $\overline{G}$ joins exactly the pairs who are strangers. A triangle in G is three people who are mutually friends; a triangle in $\overline{G}$ is three who are mutual strangers. The six-people theorem — every red/blue coloring of the edges of K_6 has a monochromatic triangle — says precisely that for *any* friendship pattern on six people, at least one of G and $\overline{G}$ contains a triangle. In Ramsey notation this is the statement $R(3,3) \leq 6$. ■

At least one of $G, \overline{G}$ has a triangle; i.e. $R(3,3) \leq 6$

The Flash

Problem 16. Königsberg landmass degrees $3,3,3,5$: is a single-stroke bridge walk possible?

SOLUTION. All four landmasses have odd degree. An Euler trail needs 0 or 2 odd-degree vertices; four exceeds that, so no single-stroke walk crossing every bridge once exists.

No (4 odd vertices > 2)

Problem 17. A connected planar graph has $V = 10$ and $F = 6$. How many edges?

SOLUTION. By Euler's formula, $E = V + F - 2 = 10 + 6 - 2 = 14$.

$E = 14$

Problem 18. A double-nine domino set (55 tiles, numbers 0–9): can they be laid in one closed ring?

SOLUTION. Dominoes are edges; a ring is an Euler circuit. Let the ten numbers 0–9 be vertices, each tile $[a \mid b]$ an edge from a to b, and each double $[a \mid a]$ a loop. Each number meets the nine others once (9 edge-ends) and its own double adds 2 more, giving degree 11 — odd — at all ten vertices. A closed ring is an Euler circuit, which requires *every* degree even (a mere trail tolerates at most two odd vertices). Ten odd vertices is hopeless, so no such ring exists.

No (all ten vertices have odd degree 11)

Proof Lab

Problem 19. Prove $V - E + F = 2$ for connected planar graphs by cycle-stripping, with a base case.

SOLUTION. Delete cycle edges until a tree remains; the alternating sum never changes. Draw the connected graph in the plane. While it contains a cycle, delete one edge of that cycle. That edge borders two distinct faces, which merge into one, so E drops by 1 and F drops by 1; the quantity $V - E + F$ is unchanged, and the graph stays connected. Repeat until no cycle remains: the graph is now a *tree*. A tree on V vertices has $V - 1$ edges and encloses no region, so it has a single face, $F = 1$. Thus

$$V - E + F = V - (V - 1) + 1 = 2.$$

Since each deletion preserved $V - E + F$, the original graph also had $V - E + F = 2$. ∎

Problem 20. Prove a connected graph has an Euler trail iff it has 0 or 2 odd-degree vertices.

SOLUTION. Trace the trail through each vertex. (*Necessity.*) Suppose an Euler trail exists. Each time the trail passes *through* an interior vertex it uses one edge to arrive and one to leave, consuming edges in pairs; so every vertex has even degree, *except* possibly the start and end of the trail, which have one unmatched edge each. Hence the number of odd-degree vertices is 0 (if the trail is closed, start = end) or 2 (open trail).

(*Sufficiency.*) Conversely, suppose a connected graph has 0 or 2 odd-degree vertices. Build the trail greedily: starting at an odd vertex (or anywhere if there are none), keep walking along unused edges. One shows by induction on the number of edges that this always extends to a trail using every edge whenever the walk gets stuck it has returned to its start, leaving a smaller even-degree subgraph on the remaining edges, which (by connectivity) shares a vertex with the walk and can be spliced in. The parity condition guarantees the walk can only get stuck at the designated endpoints. ∎

Problem 21. Prove the necessary direction of Hall's theorem, and locate the pigeonhole step.

SOLUTION. If a group is over-subscribed, pigeonhole forces a clash. Suppose a perfect matching exists, and consider any set S of people. In the matching, each person of S is paired with a distinct partner, and every such partner is acceptable to that person — so all $|S|$ matched partners lie in $N(S)$, the set of partners acceptable to someone in S. Distinct people get distinct partners, so $N(S)$ contains at least $|S|$ elements: $|N(S)| \geq |S|$. Contrapositive: if some S has $|N(S)| < |S|$, then the $|S|$ people are matched into fewer than $|S|$ partners, and by the *pigeonhole principle* two of them must share a partner — impossible in a matching. Hence Hall's condition is necessary. ∎

Problem 22. Prove every finite simple planar graph is 6-colorable.

SOLUTION. A planar graph always has a low-degree vertex; induct. First, every simple planar graph has a vertex of degree ≤ 5. If every vertex had degree ≥ 6, then $\sum_v \deg(v) \geq 6V$, so $2E \geq 6V$, giving $E \geq 3V > 3V - 6$ — contradicting the planar bound $E \leq 3V - 6$. So some vertex v has degree ≤ 5.

Now induct on the number of vertices. Remove v; the remaining graph is still simple and planar, and by the inductive hypothesis it can be properly 6-colored. Restore v: it has at most 5 neighbors, which use at most 5 of the 6 colors, leaving at least one color free for v. Coloring v with a free color completes a proper 6-coloring. The base case (six or fewer vertices) is trivial. ∎

Problem 23. Prove Gomory's theorem: removing one black and one white cell from the 8×8 board always leaves a tileable board.

SOLUTION. Build one closed tour, then let the two removals split it into even arcs. The grid graph of the full board has a *Hamiltonian cycle* — a closed tour visiting all 64 cells by side-steps. One explicit tour: go up the first column, down the second, up the third, and so on, then return home along the bottom row. Because consecutive cells of the tour are side-adjacent, they always differ in color, so the colors alternate black, white, black, white, all the way around the cycle.

Now delete the two chosen cells, one black and one white. They cut the cycle into two arcs (if the two cells happen to be neighbors on the tour, one arc is empty — an even length of zero). Walking either arc from one removed cell to the other, the colors alternate and the two ends are opposite colors, so each arc contains an *even* number of cells. Tile each arc by pairing consecutive cells two at a time: consecutive tour cells share a side, so every pair is a legal domino, and an even arc is covered exactly. Both arcs tile completely, so all 62 remaining cells are tiled. ∎

Problem 24. Prove no closed knight's tour exists on the 5×5 board.

SOLUTION. The knight-move graph is bipartite; a closed tour would be an odd cycle. A knight's move changes one coordinate by ± 1 and the other by ± 2, so it changes $i + j$ by an odd amount — every move lands on the *opposite* checkerboard color. Thus the knight-move graph is bipartite, with the checkerboard coloring as its bipartition. A closed knight's tour would visit all 25 cells and return to the start: a cycle through 25 vertices, i.e. a cycle of *odd* length. But a bipartite graph contains no odd cycle (equivalently, the 25 cells split 13–12 between the colors, and a closed tour alternating colors would need them equal). This is a contradiction, so no closed knight's tour exists on the 5×5 board.
∎

The Unpecked Chicken

Full worked solutions. Each one slices the quantity into simple indicators and adds, ignoring all dependence.

Launch

Problem 1. Expected value of a single fair die roll.

SOLUTION. Each face $1, \ldots, 6$ is equally likely, so the expectation is the average:

$$\frac{1+2+3+4+5+6}{6} = \frac{21}{6} = 3.5.$$

3.5

Problem 2. Expected number of heads in 100 fair coin flips, using indicators.

SOLUTION. One indicator per flip. Let $X_i = 1$ if flip i is heads, else 0. The number of heads is $X = X_1 + \cdots + X_{100}$, and $E[X_i] = \frac{1}{2}$. By linearity,

$$E[X] = \sum_{i=1}^{100} E[X_i] = 100 \cdot \tfrac{1}{2} = 50.$$

50

Core

Problem 3. Five hats returned at random: expected number who get their own?

SOLUTION. One indicator per person. Let $X_i = 1$ if person i gets their own hat. Each person is equally likely to receive any of the 5 hats, so $E[X_i] = \frac{1}{5}$. By linearity (ignoring the dependence between people),

$$E\Big[\sum_{i=1}^{5} X_i\Big] = 5 \cdot \tfrac{1}{5} = 1.$$

1

Problem 4. Expected value of the sum of two dice, by linearity.

SOLUTION. Let X and Y be the two die values. Each has expectation 3.5, so $E[X+Y] = E[X] + E[Y] = 3.5 + 3.5 = 7$.

7

Problem 5. Expected number of fixed points of a random permutation of $1, \dots, n$.

SOLUTION. One indicator per position. Let $X_i = 1$ if position i holds its original value i. Since each value is equally likely in each position, $E[X_i] = \frac{1}{n}$. By linearity,

$$E\Big[\sum_{i=1}^{n} X_i\Big] = n \cdot \tfrac{1}{n} = 1,$$

independent of n. **The expected number of fixed points is 1.**

1

Stretch

Problem 6. Expected position of the first Ace in a shuffled 52-card deck (4 Aces); general formula.

SOLUTION. Symmetry of the gaps. The 4 Aces split the 48 non-Ace cards into 5 groups: those before the first Ace, between consecutive Aces, and after the last. By symmetry, each of the 48 non-Aces is equally likely to fall in any group, so on average the groups are equal and the expected number of non-Aces *before the first Ace* is $\frac{48}{5} = 9.6$. The first Ace is the very next card, at expected position

$$9.6 + 1 = 10.6 = \frac{53}{5}.$$

General formula. For m "special" cards among n total, the same gap-symmetry gives expected first-special position $\frac{n-m}{m+1} + 1 = \frac{n+1}{m+1}$. (Check: $\frac{52+1}{4+1} = \frac{53}{5} = 10.6$.)

$\frac{53}{5} = 10.6$; **general** $\frac{n+1}{m+1}$

Problem 7. Expected number of boxes to collect all $n = 6$ coupons; why the last dominates.

SOLUTION. Sum the stage-waits. Break the collection into stages by how many distinct coupons you hold. When you have k distinct coupons, a box gives a *new* one with probability $\frac{n-k}{n}$, so the expected number of boxes to advance to $k+1$ is $\frac{n}{n-k}$ (the mean of a geometric distribution). By linearity, the total expected boxes is

$$\sum_{k=0}^{n-1} \frac{n}{n-k} = n\left(\frac{1}{n} + \frac{1}{n-1} + \cdots + \frac{1}{1}\right) = n\,H_n.$$

For $n = 6$: $6H_6 = 6 \cdot \frac{49}{20} = \frac{147}{10} = 14.7$ boxes. Why the last coupon dominates: the final stage (going from 5 to 6 coupons) has success probability only $\frac{1}{6}$, so it alone averages $\frac{6}{1} = 6$ boxes — more than a third of the total — since the last coupon is rare to draw.

$6H_6 = 14.7$

Problem 7b. n chickens in a circle each peck a random neighbor. Expected number pecked by both neighbors at once; check the three counts sum to n.

SOLUTION. One indicator per chicken, then linearity. Focus on a single chicken C. It is pecked by its left neighbor only if that neighbor pecks *right* (toward C), probability $\frac{1}{2}$; similarly its right neighbor pecks C with probability $\frac{1}{2}$, and the two neighbors choose independently. So:

$$P(\text{pecked by both}) = \tfrac{1}{2} \cdot \tfrac{1}{2} = \tfrac{1}{4}, \quad P(\text{un-pecked}) = \tfrac{1}{2} \cdot \tfrac{1}{2} = \tfrac{1}{4},$$

and $P(\text{pecked exactly once}) = 1 - \frac{1}{4} - \frac{1}{4} = \frac{1}{2}$. By linearity over all n chickens, the expected number pecked twice is $\frac{n}{4}$, un-pecked $\frac{n}{4}$, and pecked-once $\frac{n}{2}$. These sum to

$$\frac{n}{4} + \frac{n}{2} + \frac{n}{4} = n,$$

exactly as required, since every chicken falls into one of the three categories. (For $n = 100$: 25 pecked twice, 25 un-pecked, 50 pecked once.) ■

$\frac{n}{4}$ pecked twice; the three counts sum to n

The Flash

Problem 8. Expected number of flips to the first head (fair coin).

SOLUTION. The number of flips until the first head is geometric with success probability $p = \frac{1}{2}$, whose mean is $\frac{1}{p} = 2$.

2

Problem 9. Expected number of both-heads pairs in 10 flips.

SOLUTION. One indicator per pair. There are $\binom{10}{2} = 45$ pairs of flips. Each pair is both-heads with probability $\frac{1}{2} \cdot \frac{1}{2} = \frac{1}{4}$, so by linearity (the pairs overlap, but linearity doesn't care),

$$45 \cdot \tfrac{1}{4} = \tfrac{45}{4} = 11.25.$$

11.25

Proof Lab

Problem 10. Prove $E[X+Y] = E[X] + E[Y]$ from the definition; why no independence is needed.

SOLUTION. Sum over outcomes. By definition, expectation sums the value times probability over all outcomes ω of the sample space:

$$E[X+Y] = \sum_{\omega} (X(\omega) + Y(\omega))P(\omega) = \sum_{\omega} X(\omega)P(\omega) + \sum_{\omega} Y(\omega)P(\omega) = E[X] + E[Y].$$

The splitting of the sum is just the distributive law; at no point did we multiply probabilities of X and Y together, so *no independence assumption is used*. Linearity holds for any random variables on the same sample space, however they depend on each other. ∎

Problem 11. Derive the coupon-collector expectation nH_n, justifying the $\frac{1}{p}$ wait.

SOLUTION. Each stage is a geometric wait. Suppose you currently hold k distinct coupons. Each new box is a fresh independent draw, uniform over the n coupons, so it shows a coupon you lack with probability $p_k = \frac{n-k}{n}$. The number of boxes you buy until the first success is therefore a geometric random variable with parameter p_k.

Why a geometric wait averages $\frac{1}{p}$: if W is the number of trials until the first success with per-trial probability p, then conditioning on the first trial gives $E[W] = 1 + (1-p)E[W]$ (a failure resets the count after one wasted trial), so $pE[W] = 1$ and $E[W] = \frac{1}{p}$.

Thus the expected boxes to pass stage k is $\frac{1}{p_k} = \frac{n}{n-k}$. Summing the stages by linearity,

$$E[\text{total}] = \sum_{k=0}^{n-1} \frac{n}{n-k} = n \sum_{j=1}^{n} \frac{1}{j} = n H_n.$$

∎

www.ingramcontent.com/pod-product-compliance
Lightning Source LLC
LaVergne TN
LVHW060454061126
851857LV00016B/1820
9798906210081